SCIE

For the Men and Women who aspire to the highest sustainable level of leadership without destroying the rest of your life

Mental Skills Leadership *for* Profession & Life

Mental Skills:
Goals   Imagery   Relaxation   Refocus

Number 1 Cause of Failure in Leadership and in Life–Inability to REFOCUS

Plus: The Purple Sweater
Sam L Hughes PhD

Copyright © 2013 Sam L Hughes PhD
All rights reserved.
ISBN: 0615784542
ISBN 13: 9780615784540

DEDICATION

To the excellent role models who serve in leadership and who display great character in life and profession

DEDICATION

Acknowledgement

I wish to acknowledge the love, sacrifices and support of my wife and children throughout my life's journey. Also to my mother, father, and grandparents who raised me in an environment of non prejudice regarding gender, race, and nationality.

In quest of my doctoral degree these individuals of great character had a positive impact on my life.

Leon Griffin, PhD; John Rinaldi, PhD; David Scott, EdD; Todd Seidler, PhD; Armond Seidler, PhD

Contents

Dedication	iii
Acknowledgement	v
Preface	ix
Part I: Science and Philosophy of Your Leadership	15
Chapter 1: An Individual: You	17
Chapter 2: Visionary Goal: Yours	63
Chapter 3: .Motivate Team Members	85
Part I: Summary	133
Part II: Science and Art of Mental Skills	135
Chapter 4: Goals	141
Addendum: The Human Brain	189
Chapter 5: Imagery	195
Chapter 6: Relaxation (Alpha Brain Waves)	231
Chapter 7: Refocus	255
Part II: Summary	283
Mental Skills Leadership for Profession and Life: Summary	287
References	293
Glossary	297

Preface

Mental Skills Leadership for Profession and Life is a book for men and women who aspire to the highest level of leadership with sustained greatness and for those who believe *leadership* has a role and a duty of positive modeling for the benefit of those they serve and for all who observe. Philosophy and science of the mind are applicable in leadership and to one's personal life, including life crisis. It is a book not just for those at the highest level of leadership, but also for those who participate throughout every level of leadership in every profession.

The need for excellence in leadership has always been necessary. However, today's challenges in all professions are greater than in the past. In cultures worldwide, vital and complex issues are at their highest level. Cultural problems and deviance filter into leadership's structure. These issues challenge leaders to elevate leadership to its highest sustainable level. The need to understand and maintain the balance of your character with profession is critical. Failure of this understanding will eventually lead to your leadership imploding, causing tragic consequences to yourself and to those you serve.

Mental Skills Leadership (MSL) provides a method, a strategy, for allowing today's leaders and future leaders to step back and evaluate their critical balance, **What Am I** (your profession) with **Who Am I** (your character). Because of today's dynamic and vital issues, *MSL* places major emphasis on evaluating and maintaining balance of your professional and personal life.

When your leadership fails to periodically assess this critical balance, the negative results will range from major to catastrophic. Consequences of this imbalance are witnessed almost daily on every level of leadership. Far too often we witness leaders and team members' personal lives imploding because of failure to maintain the balance of personal character with profession.

It is incumbent upon each one of us to continually remind ourselves that we must take charge and direct the course of leadership. "Leaders with honest self-awareness of their character typically know their limitations and strengths, and exhibit a sense of humor about themselves. They exhibit a gracefulness in learning where they need to improve and welcome constructive criticism and feedback." (Goleman, Boyatzis, and McKee, 2004, p. 254) When appropriate, *MSL* is laced with light hearted humor allowing you to observe possible benefits of self-humor.

Mental Skills Leadership encourages your personal thoughts and your self-evaluation. *MSL*, not only presents the qualities necessary for elevating leadership to its highest sustainable level, it also presents how you may achieve each necessary quality with proper balance.

Regardless of leadership style or leader personality, *MSL* is an enabler for elevating your leadership to the highest sustainable level. However, an alarm must be sounded concerning two tenets of toxic leadership which poison team environment:

1. Hubristic Leader
2. Insecure Leader

Mental Skills Leadership addresses both of these issues.

MSL follows the linear structure of the following leadership definition.

An Individual *with a* **Visionary Goal** *that has the ability to* **Motivate Team Members** *towards* **Accomplishing Your Visionary Goal**

*Bold words in this definition will be discussed in subsequent chapters.

Regardless of profession, *Mental Skills Leadership* refers to those you serve as Team Members. Two demographics within the definition of leadership:

1. Leader (you) 2. Team Members (those you serve)

PART I: SCIENCE AND PHILOSOPHY OF YOUR LEADERSHIP

- Provides you the opportunity of self-evaluation for developing and exploring your philosophy of life (character) and your philosophy of profession (purpose).
- Provides an enjoyable, thought provoking, practical, important experience for creating a strategy from which you may elevate your leadership to its highest sustainable level and for those aspiring to become great leaders.
- Provides a process for evaluating your critical balance, *What Am I* with *Who Am I*. Everything in your leadership starts with your character ... EVERYTHING.

PART II: SCIENCE AND ART OF MENTAL SKILLS

- Provides a menu of the most recently proven mental techniques from which you may pick and choose to achieve the most productive, engaged team members in pursuit of accomplishing your visionary goal. Science is the knowledge of the Mental Techniques: *Goals, Imagery, Relaxation, Refocus.* Art is the application of each mental skill.

- Furthermore, **Part II Addendum** provides basic information of the human brain for better understanding the science and art of mental skills.

YOUR LEADERSHIP: (DEVELOPS TWO LEADERSHIP GOALS)

- Visionary Goal: Your vision for the future. In the process of elevating leadership to a higher sustainable level all efforts, by you and your team members, are to successfully accomplish the ultimate target, your visionary goal. (Chapter 2)

- Performance Goal: Your personal strategy to accomplish your Visionary Goal

YOUR LEADERSHIP: (ASSISTS DEVELOPMENT OF TWO TEAM MEMBERS GOALS)

- Destination Goal: Team members develop their *Destination Goal* (Long term goal for targeted outcome) Team member destination goal must compliment the accomplishment of your leadership's visionary goal.

- Performance Goal: Team members develop their *Performance Goal* (Short term goal) Strategy to accomplish a desired upcoming performance.

LINEAR PROGRESSION OF THE FOUR GOALS:

- **Leadership**: (1) Visionary, (2) Performance
- **Team Member:** (3) Destination, (4) Performance
- **Mental Skill Chapters 4, 5, 6, 7:** (Goals, Imagery, Relaxation, Refocus)

Each mental skill is divided into two major categories:

- **Science:** *Knowledge* of the mental skill
- **Art:** *Application* of the mental skill

The philosophies, as well as the science and art of mental skills presented in *MSL,* enable you to craft the perfect team culture. Elevating your leadership to a higher sustainable level requires bringing hope, enthusiasm, and passion each day for team members. *Mental Skills Leadership* presents a systematic daily process for developing a motivational environment that permeates throughout your organizational structure. When leaders attend only casual or random concern for science of the mind in their team culture, the full potential of team will be limited.

The challenge is not only to pursue excellence but to do so without destroying the rest of your life. T.Orlick, 2007

AN **INDIVIDUAL** WITH A **VISIONARY GOAL** THAT HAS THE ABILITY TO **MOTIVATE TEAM MEMBERS** TOWARDS ACCOMPLISHING **YOUR VISIONARY GOAL**

PART I

SCIENCE AND PHILOSOPHY OF YOUR LEADERSHIP

Be more concerned with your character than your reputation, because your character is what you really are, while your reputation is merely what others think you are.
JOHN WOODEN

PART I: IS DEDICATED TO YOU AND YOUR LEADERSHIP

The following discussion allows you the opportunity to exercise fair and honest self-evaluation of your character and your leadership. Every individual who has had a career of successful leadership and sustained this level understood and maintained their critical balance, **What Am**

I with **Who Am I**. Failure to maintain the critical balance increases the possibility of consequences which may not be repairable. You are given the opportunity to evaluate your critical balance. Also presented are strategies for developing a motivational environment for your team members.

Part I: Follows the linear sequence of bold words in the definition of leadership
Chapter 1: An Individual (You)
Chapter 2: Visionary Goal (Yours)
Chapter 3: Motivate Team Members (Those You Serve)
Part II (Ch. 4, 5, 6, 7) the finality of bold words in Leadership definition: **Accomplishing Your Visionary Goal**

The qualities of leadership excellence transcend all professions and personal life. **Part I**, when applicable presents a broader focus on different professions. It is our contention when we step back and view leadership through the lenses of other professions such as military, sport, business, education, religion, politics, entertainment etc., it will bring an enhanced clarity when returning a narrow focus to your professional leadership.

Throughout *MSL* your ability to discover ideas from another profession and apply these ideas to your profession will be beneficial. Your ability to perform this mental function is referred to as transference. One of the major benefits of transference is longevity of your applied discoveries.

Chapter 1:

An Individual: You

INTRODUCTION

AN **INDIVIDUAL** WITH A **VISIONARY GOAL** THAT HAS THE ABILITY TO **MOTIVATE TEAM MEMBERS** TOWARDS ACCOMPLISHING **YOUR VISIONARY GOAL**

Of all the evaluations we make in life none is more critical and important than those we make about ourselves. (Martens)

Chapter 1 addresses the first bold word **(Individual)** in the definition of leadership. Furthermore, this chapter places major emphasis on how to successfully maintain your critical balance, *What Am I* with *Who Am I*, while elevating your leadership to its highest sustainable level. Without this balance one of two results will eventually occur: your leadership will self- destruct or your personal life will implode. In leadership and in your personal life your character is the Alpha and the Omega. *Philosophy of Profession* and *Philosophy of Life* are presented for your personal discovery and self-evaluation.

CHAPTER 1 AN INDIVIDUAL (YOU): MAJOR TOPICS

WHO AM I

SELF-EVALUATION OF YOUR CRITICAL BALANCE:

- What Am I
- Who Am I

DEVELOPING YOUR PHILOSOPHIES:

- Your Philosophy of Life
- Your Philosophy of Profession

Chapter 1 Glossary: Hubris, Insecure Leader, Philosophy of Life, Philosophy of Profession, Secure Leader, Stolen Identity, What Am I, Who Am I

An Individual: You

AN **INDIVIDUAL** WITH A **VISIONARY GOAL** THAT HAS THE ABILITY TO **MOTIVATE TEAM MEMBERS** TOWARDS ACCOMPLISHING **YOUR VISIONARY GOAL**

Chapter 1 is the foundation chapter of *Mental Skills Leadership* (*MSL*). With an open mind and a sincere attempt to personalize this information, only then will the remaining chapters provide full discovery for elevating and sustaining successful leadership in your profession and in your personal life. The importance of responsibility you inherit in leadership is immeasurable. *Mental Skills Leadership* is designed to assist improvement of your personal style of leadership to its highest sustainable level.

You are presented opportunities to evaluate your critical balance of *What Am I* (Profession) with *Who Am I* (Your Character). You are also given the opportunity for self- evaluation and personal opinions in developing *Your Philosophy of Life* and *Your Philosophy of Profession*.

In the process of enhancing your present leadership style to a higher level and for those aspiring to become great leaders, you must ask yourself certain questions with answers discovered deep within your heart and soul. Seeking these personal answers is paramount for great leadership. You will find enjoyment, pleasure and surprise in developing a solid foundation for pursuing your highest level of sustainable, successful leadership.

A major quality separating your leadership value (bad, good, or great) will depend upon how you respond to your mistakes. Your leadership is a perfect mirror or reflection of your character. It is your character which determines sustainability of success.

Great leaders in all profession and life have three like qualities. Each of these qualities is systematically developed; they are not inherent. Throughout each chapter *Mental Skills Leadership* pursues the following necessary qualities for your sustainable success.

- **Great Self-Confidence** (belief in competence to achieve)
- **Do not Fear Failure** (make failure a learning process for future success)
- **Developed Ability to Refocus** (when things don't go as planned mentally getting back on track)

*Number 1 Cause of Failure in Leadership and Life: *Inability to Refocus*

On several occasions *MSL* affords you the opportunity to provide your personal thoughts. In doing so we highly recommend you develop a personal code such as the following example in which you are presented the opportunity to choose <u>any two consecutive numbers.</u> Example of Coding: (0 1) 2 3 <u>4 5</u> 6 7 "8 9" In this example you provided three answers (0-1), <u>4-5</u>, "8-9", only you know which of the three is the correct answer.

Personal Assessment: Evaluate your leadership qualities as you perceive them to be. (Choose <u>any 2 consecutive numbers)</u>

0-1 none,
5-6 average,
8-9 great

I believe my level of Self-confidence in my Leadership to be:

0   1   2   3   4   5   6   7   8   9

I believe my level of Self-confidence in my Personal Life to be:

0   1   2   3   4   5   6   7   8   9

I believe my level of not Fearing Failure to be:

0   1   2   3   4   5   6   7   8   9

I believe my level of Ability to Refocus to be:

0   1   2   3   4   5   6   7   8   9

I believe my level of Ability to properly attend to my Mistakes:

0   1   2   3   4   5   6   7   8   9

"Good leaders are just as close to bad leaders as they are to great leaders." (Lynch & Chungliang, 2006, p. 4)

SELF-EVALUATION OF YOUR CRITICAL BALANCE

WHAT AM I WITH WHO AM I

Your professional leadership accomplishments, *What Am I*, are terminal, no matter how successful. Over time your accomplishments will lose original meaning. Your character, *Who Am I*, is eternal with lives you have touched. These qualities pass from generation to generation. The balance of your profession with your character determines sustainability of success.

In your leadership position, for the sake of self-survival in your professional and personal life, you must routinely evaluate this critical balance: *What Am I* with *Who Am I*. Maintaining your balance is not a luxury; it is a necessity.

IMPLODING LEADERSHIP: (DESTRUCTION FROM WITHIN)

Recently one of the saddest tragedies in modern-day leadership occurred in the United States at a Division I University's highly successful football program. Leadership failed to systematically evaluate the critical balance, *What Am I* with *Who Am I*. Not only was this implosion devastating to leadership, but also hurtful to family members, team members, children, and to the legacy. It is possible you may recall implosion in different professions such as military, business, religion, entertainment, politics and others.

When your leadership dedicates all efforts and concerns entirely to your profession and fails to consider those outside of your profession, you are allowing the possibility of your legacy to be tarnished. Concern for those other than your profession should be included in evaluation of *Who Am I*. Elevating your leadership to its highest sustainable level necessitates a sincere systematic self-evaluation of *Who Am I*. On each occasion during the process of evaluating your team members, you should also evaluate your critical balance using a 0-9 scale or a similar measurement.

In leadership and in life we are comprised of two major tenets:

- **What Am I** (Your Professional Leadership Position) The *What Am I* is easily answered since it is primarily a constant. *What Am I* is your committed purpose to your professional leadership position.

- **Who Am I** (Your Character) Not so easy ... is the answer to the more complicated question "*Who am I?*" *Who Am I* is not a constant; it is indeed a dynamic variable, always evolving. *Who am I* is all about your personal character, a living, growing entity flexible and ever changing with each thought and decision circumstance.

Inevitable Self-Destruction: Stressing the balance of *What Am I* with *Who Am I*

In leadership one may make tens-of-thousands correct decisions, then make one critical incorrect decision, negating or tarnishing all of the positive accomplishments. Most often this grave mistake is a result of failing to keep your finger on the pulse of your critical balance. This is the most important, consistent, tenacious daily challenge for individuals on all levels of leadership. Keeping this balance requires a disciplined focus.

Caution: Without this balance the road to self-destruction becomes a super highway with no speed limits and very few available exit ramps. This imbalance is witnessed everyday on local, national and international news where individuals in personal life and in their profession have self-destructed. The many lives affected and the harm caused is immeasurable.

Hubristic attitudes and rationalization of wrong to right greatly challenge your critical balance in leadership. Rationalizing wrong to right has never worked, and never will, in the process of elevating and sustaining successful leadership. It is important your self-evaluation be a sincere, honest and fair cognitive process. Not being fair to yourself is counterproductive.

WHAT AM I VS WHO AM I

(Applied)

Critical Balance in Leadership and Life

As a sport scientist I helped a friend who was the head football coach at a Division I University in the Southeastern Conference (SEC). The SEC is often argued as one of the toughest, if not the toughest football conference in the United States. I had the privilege of implementing a two year Mental Skills Leadership Program (MSL) in the sport of football. The first year, the MSL Program helped produce the most victories for my friend in his tenure. The second year the football program was undefeated in the nation.

Addressing the 112 member squad for the first time, I informed the student/athletes the topic was "You ... each one of you." Continuing, I related to the team, each one of us is actually made up of two different entities:

- **What Am I** (your profession: Student/Athlete)
- **Who Am I** (your character)

I informed the squad I knew what each one of them was in spite of the fact I had never met any of them. "YOU are highly talented student/athletes at one of this nation's greatest universities. I do not know *Who* each one of you is. Chances are you do not either. You see, *Who Am I* is all about your character—a living, dynamic variable. To be honest with you, I really do not know *Who Am I*."

I explained to these fine young men, "There are no heroes in this room because of your performance on the playing field. You may be a star, a super star, player of the game, conference player of the year, or even a Heisman Trophy winner; but that does not make you a hero. Often times sport announcers or sport writers call you a hero. However, this is not true; the rules of team sport do not permit heroes on the playing field." Having made these comments, I wondered how these wonderful young men would react. Their relief was visible from this unintentional, unfair characterization of their sport performances.

"Real heroes are the men and women of the military. Real heroes are first responders. Would you like to shake hands with the greatest hero or would you like to give a hero a hug? You see, your greatest hero is the person(s) responsible for your gluteus maximums being here right now. The moms and dads, the grandmothers and grandfathers, and those who helped raise you are the greatest heroes. These heroes have made sacrifices you may not know about until your later years in life; some sacrifices you will never know. Heroes are those who are willing to give of themselves on a regular basis to a cause greater than themselves."

As of this writing 12 of these collegiate student/athletes are starters in the United States National Football League (NFL), one of which has two Super Bowl rings. Many of these highly talented athletes are good role models with excellent character. They have successfully maintained their critical balance, *What Am I* with *Who and I*. I am incredibly proud of these young men. Their good character has survived in a profession in which *Who Am I* is challenged daily.

An important question: Presently how successful are you at keeping your critical balance, *What Am I* (profession) with *Who Am I* (your character)? If you were to give yourself an evaluation of this critical balance, 0-1 horrible and 8–9 great, what score do you believe is fair? (Choose any two consecutive numbers-remember to code your evaluation.)

0   1   2   3   4   5   6   7   8   9

EVALUATING LEADERSHIP IN POLITICS

Profession of Politics

Consider for example the profession of politics. Has the profession of politics over time developed a culture of transparent, in-your-face deviance? Are lying, cheating, corruption, buying votes, and thinking of one's self-interest character traits of great leadership? When it is necessary to make hard decisions and one hides under their desk, is this character? The scope of these deviant characteristics may be witnessed from local public offices to the highest offices in the land in many free nations.

When events get tough and one throws constituents (team members) under the bus, is that great leadership? Publicly trashing the messenger because one doesn't like the message, is that *Who Am I?* Is it possible this type individual is simply a narcissist with a hubristic attitude occupying a leadership position?

Caution: Just because one occupies a leadership position does not make one a leader.

Illinois Politics

Many Illinois citizens believe for each vote cast for a certain political party at least five dead people in a northern and southwestern county vote for the other party. Recently, my physician related the following story. "After my father died, for the next five years he voted a straight ticket. I had to personally go to the election board and insist my father's name be withdrawn as an eligible voter."

Three Leadership Questions for Politicians: Ask any national, state, local, politician the following three questions, odds are many will be mystified or simply dumb-founded by the questions:

- What is your visionary goal for your nation?
- What is your philosophy of politics?
- What is your philosophy of life?

These questions are applicable to leadership in all professions in every free nation. In reading *Mental Skills Leadership* you will be given the opportunity to develop your personal response to the three previous vital questions.

Questions: Is It Possible? (Your Decide)

- Is it possible politics on many levels have developed a culture where deviant norms are accepted and Who Am I no longer has importance?
- Is it possible the culture of no character exists in other professions? What about your profession? What about in your own life?
- Is it possible no character or bad character exists throughout many national cultures?
- Is it possible your leadership role in your profession inherits a duty for role modeling excellence?
- Is it possible your leadership role includes providing life skills for your team members?

The importance of this critical balance applies to all professions. The final question comes down to this: When each of us looks in a mirror what type of character do we see? Do you see an acceptable balance? *What Am I* balanced with *Who Am I* requires an honest, fair, systematic self-evaluation. This self-evaluation applies to professional and personal life. When leaders accept responsibility and duty to display good character, the benefits filter into one's national culture. The same is true when leadership fails to provide character excellence.

On occasion high profile figures, whether leaders or team members, will state, "I am not a role model." This statement is simply not true. It is an excuse for bad behavior or lack of character. Any individual with a high public profile is a role model. Your public and private behavior is a mirror of your character; good or bad-the *Who Am I*.

WHAT AM I

Your Profession

Regardless of profession, there are many similarities and very few differences in leadership. The militaries who defend freedom and first responders are professions in which a major difference exists. In these professions leadership and strategy may be executed perfectly and still the ultimate price may be paid. The military of free nations and first responders are excellent role models, displaying great character.

However, military personnel are just as susceptible as others in failure to maintain the critical balance of *What Am I* with *Who Am I*. Science of human behavior indicates everyone is a candidate for failure to balance profession with character. More simply stated, it is the present stage of our brain's evolution or as some suggest the brain's devolution.

Leadership in all professions has two demographics:

- Leader (You)
- Team members (Those you serve)

Questions:

- Within your profession can you think of an individual who displays or displayed great leadership?
- Within your profession can you think of an individual who displays or displayed poor leadership?
- In your personal opinion, what separates the great leadership from the poor leadership you chose in the previous two questions?

In cultures worldwide it is not uncommon to hear or read about movie stars, religious leaders, military leaders, military personnel, coaches, athletes, politicians, entertainers, or business men and women whose lives have become extremely complicated or completely out of control. Over time and subconsciously, *What Am I* has completely overridden and destroyed *Who Am I*. Every individual on the third planet from the sun is susceptible to this destructive imbalance, no exceptions. This important balance is also crucial in personal relationships.

WHAT AM I CONSUMES WHO AM I

Stolen Identity

The imbalance of *What Am I* with *Who Am I* is often the cause of troubled personal relationships. Either knowingly or unknowingly one of the partners over time completely consumes the *Who Am I* of their counterpart. In such instances this imbalance increases the destructibility of a successful relationship.

This same imbalance is true in professional leadership with the exception of the source. The source of destruction in professional leadership is you. The source of destruction in personal life is either you or your counterpart. Without a systematic self-evaluation of your critical balance the identity of *Who Am I* will be harmed: Stolen Identity is painful and often catastrophic whether in your profession or personal life.

A lesson from the movie Cool Runnings

"I made winning my way of life and when you do that you have to keep on winning at any cost."

This statement is the result when *What Am I* (winning) completely overrides and destroys *Who Am I* (character).

The scenario: Coach Irv, an American Olympic athlete, had won two gold medals in bobsledding. However, he was caught cheating, had his metals revoked, and was given a lifetime suspension from competing as an athlete in Olympic competition. After several years had passed, this disbarred athlete made Jamaica his home away from home. He spent most of his time in the local bars.

Four young talented Jamaican athletes wanted to participate in the upcoming Winter Olympics. All the Jamaican events had been filled with the exception of the bobsled team. These four young Olympic hopefuls needed a coach.

Finally, they were able to convince Coach Irv to be their coach in preparing for the Jamaican bobsled team. Having found an old rickety bobsled stashed in a barn, they practiced religiously until they successfully made it to the final trials in the Olympics.

The evening before the final run Coach Irv stopped in on the captain of the team at his Olympic apartment to see if he wanted to go get a bite to eat. After the Captain politely declined, Coach started to leave the room.

Before Coach Irv could exit the apartment the young athlete make the following comment, "Coach, I'd like to ask you something, but you don't have to answer if you don't want."

"You want to know why I cheated, don't you?" asked the coach.

"Yes I do."

"That's fair enough. You see, **I made winning my way of life, and when you do that, you have to keep on winning at any cost.**"

"But coach, you had it all; you had two gold medals."

The coach gave pause for thought, "You see a gold medal is a wonderful thing, but **if you are not enough without a gold medal, you will never be enough with it.**"

The captain responded, "How will I know if I am enough?"

"You'll know when you cross the finish line." The coach was trying to tell his young athlete there are two ways to cross the finish line: Cross the finish at any cost or cross the finish line with character.

"I made winning my way of life, so I had to keep winning at any cost." Did the *What Am I* (Gold Medal Winner) completely consume and destroy the *Who Am I* (Character)? The lesson from *Cool Runnings* is especially applicable to those of us who have accepted leadership positions. Leaders and team members have two choices in role modeling for those they serve and for those who observe: bad role model or good role model.

IMBALANCE

Catastrophic Result

In the early 1990's I served on a coaching staff with a wonderful and extremely talented young coach, then a tragedy occurred in this man's life. He had been highly successful for several years; he was liked and respected by his players, coaches, and fans. Following this success his teams fell upon hard times. For one and one third season his team failed to win a game. Then the shocking news ... the coach ended his life leaving behind a wonderful wife and three beautiful young children.

Remember the words of the disgraced Olympic coach: *I made winning my way of life, so I had to keep winning at any cost.* At this moment how aware of and how successful are you with the balance

What Am I with *Who Am I?* In your leadership are you pursing success at any cost?

In every profession it is possible a culture of deviance may evolve. Cheating, lying, corruption, revenge, rationalizing wrongs as rights, cronyism, nepotism, and winning at any cost are human behavioral traits threatening good character and great leadership. Regardless of profession, regardless the level of leadership, regardless the level of performance ability; these deviant character traits of human behavior are given fertile ground when *Who Am I* is totally devoured by *What Am I*.

Leadership Character:

How your character defines deviance will define the boundaries for your team members, providing what is acceptable and what is not acceptable. Deviance is the ultimate factor in determining the difference between good and bad character.

Negative behavioral traits are capable of sneaking into personal character. In self- evaluation of your character it is vital you be brutally honest while being fair to yourself. The inescapable day will come when you cross the finish line in your profession. When you look back at your leadership, what kind of character will you see? What kind of character will your team members have seen? What will be your legacy? What would you like your legacy to be?

Maintaining proper balance of *What Am I* with *Who Am I* is a mental process requiring self-discipline. Keeping this balance in elevating your leadership to its highest sustainable level is critical in leadership and in life.

In keeping the proper balance, the lyric in one of Jimmy Buffett's songs (2004) makes this point, *"life is complicated, just don't let it drive you nuts."* Keep a balance of the *What Am I* with *Who Am I*. If you don't keep this balance it will ... drive you nuts.

WHO AM I

Your Character

Nearly all individuals can withstand adversity, but if you want to test a man's character, give him power.
-ABRAHAM LINCOLN

Since the beginning of mankind each individual who strives to achieve has either consciously or subconsciously sought to answer the question, *Who Am I?* Only by honestly looking into ourselves may we find our true character. Once you have completed self-evaluation, you are now in a position to embark on a personal journey of elevating leadership to a sustainable level. You may be a great leader, but it is your character that will sustain your higher level of greatness.

HIGHLY SUCCESSFUL AND SUSTAINABLE LEADERSHIP

Legacy: Portrait of Lasting Character

I have personally followed Jerry Sloan's basketball career. While in high school, I played point guard for Carmi, Illinois; Jerry played guard-forward for McLeansboro. He handed us two of our losses, while we compiled 20 victories. In high school Jerry was unstoppable. It truly was inspiring for me to be on the same floor with him. Jerry Sloan was and is highly competitive, intense, focused, and fair. Even in high school his excellent character was obvious.

Jerry Sloan's Professional Leadership: *What Am I*

He was the only coach in history to win 1000 games with one club, the Utah Jazz; 15 consecutive playoff appearances; 1997 and 1998 lost finals to Michael Jordan and the Chicago Bulls; and the longest tenured head coach in American major league sport: Jerry Sloan retired Feb. 10, 2011.

These wonderful accomplishments of Jerry Sloan, *What Am I*, will have lost importance by the next generation. Professional accomplishments lose meaning over time and continue to fade. (*What Am I* accomplishments are terminal.)

Jerry Sloan is one of the greatest and most respected in NBA history.
NBA COMMISSIONER

Jerry Sloan's Character: *Who Am I*

The great character Jerry Sloan has demonstrated throughout his life is eternal. All the individual lives he touched and all who have observed will miss this great leader. I have personally carried a memory of his excellent character throughout my professional life of 38 years.

In our college careers Jerry enrolled at Evansville University, Evansville, Indiana, to play for the Purple Aces. Having transferred from another university, he had to sit out for a determined time by the National Collegiate Athletic Association (NCAA). I was also enrolled at Evansville University.

During this interim Jerry and I were playing in a downtown league, but once again on opposite teams. The following situation occurred, which I have kept with me in developing and maintaining my personal character. Our shooting guard's shot rebounded off towards Jerry (6'6"), standing at the free throw line waiting for the ball. Unknown to Jerry, I (6'0") was directly behind him. As the rebound approached I elevated myself over Jerry's back, placing my right hand on his shoulder and using my left hand, tipped the rebound back to the goal hitting nothing but net.

Jerry immediately turned around, waiting for my feet to hit the floor. Keeping in mind I had just fouled Jerry; my first thought was, *he may lay me out right here.* However, that is not what happened. Instead, Jerry complimented me on a great play; this coming from an athlete who is highly talented, intense, very competitive, and a man of great character.

Most likely Jerry Sloan will not recall this incident; this is simply who he is. I certainly have remembered, and I have tried to pass this excellent character quality on to other lives I have touched. Your good Character, *Who Am I*, is eternal. Not only is good character eternal, so too is bad character. Failing to evaluate the critical balance, *What Am I* with *Who Am I*, will tarnish your leadership accomplishments and those of your team members.

Systematically evaluating your character will enhance sustainability of success. Today in all professions, some in leadership positions have sold themselves the false idea that character has no benefits. It is possible, and likely, one may find success with bad character in leadership, but carefully observe its sustainability. Jerry Sloan's professional leadership is a true example, an icon of good character sustaining success at its highest level.

SELF-EVALUATION OF YOUR CHARACTER

Great Leadership begins with Good Character

Character Exercise:

In search of *Who Am I*: Looking into Ourselves

Evaluate each trait: 0-1 = No Importance, 5-6 = Average, 8-9 = Great Importance

In your opinion how important is each character trait: (Choose any two consecutive numbers)

Loyalty

0   1   2   3   4   5   6   7   8   9

Non bias

0   1   2   3   4   5   6   7   8   9

Empathy

0   1   2   3   4   5   6   7   8   9

Fairness

0   1   2   3   4   5   6   7   8   9

Passion

0   1   2   3   4   5   6   7   8   9

Self-confidence

0   1   2   3   4   5   6   7   8   9

Commitment

0   1   2   3   4   5   6   7   8   9

Trustworthy

0   1   2   3   4   5   6   7   8   9

Confidentially

0   1   2   3   4   5   6   7   8   9

Courage

0   1   2   3   4   5   6   7   8   9

Caring

0   1   2   3   4   5   6   7   8   9

Composure

0   1   2   3   4   5   6   7   8   9

Self-discipline

0   1   2   3   4   5   6   7   8   9

Determination

0   1   2   3   4   5   6   7   8   9

Other

0   1   2   3   4   5   6   7   8   9

Other

0   1   2   3   4   5   6   7   8   9

Given the opportunity to choose two of your strongest character traits, what would you choose? 1. _____ 2._____ If you were to choose a character trait to improve, what would you choose? _____

COURAGE AND PASSION IN LEADERSHIP

A Global Company

Home Depot

In a nationally televised interview in 2012, the co-founder of Home Depot Ken Langone was questioned intensively about successful leadership. Mr. Langone consistently used two words in describing leadership qualities: Courage and Passion. He also made the following statement,

"Business that doesn't have leaders with courage, don't make it; leadership requires courage." Ken Langone also talked about the values his parents instilled in him. His comment applies to leadership in every profession.

Immediately after Mr. Langone left the set, the business commentator made the following statement about this great and highly successful leader, "He is one of the most generous and humble guys around." Good character benefits vary with each individual's leadership.

CHARACTER

West Point Military Academy

In the United States of America, West Point Military Academy places major emphasis on character, *Who Am I*, in developing great leadership. It is the Academy's deep-seeded belief; the number one quality for leadership is good character. At West Point should a cadet be accused of cheating or sexual harassment and the accusation confirmed, the cadet will be expelled immediately. On the other hand should the accusation be proven false, the accuser will be expelled immediately.

I had the honor and privilege of accepting and then completing a Doctoral Internship in Leadership at West Point Military Academy. While interviewing Col. Csoka, the creator, developer, and director of the Performance Enhancement Training Program, I asked the Col. the following question: What is the difference between good leaders and great leaders? Without hesitation he replied, "Good leaders get the job done. Great leaders get the job done and lead with passion. Great leaders have self-confidence and passion in leading their soldiers (team members) in accomplishing their mission (goal)."

Major emphasis is placed on moral character. Top brass at West Point proudly and eagerly admit they work with the finest sons and daughters their nation has to offer.

> *The number one quality for leadership is character and leadership is all about people.*
> GEN. NORMAN SCHWARZKOPF

DALLAS MILLIONAIRE

A well known, self-made Dallas millionaire developed a highly successful company in Plano, Texas. Upon interviewing potential employees he would ask the following question, "Are you being faithful to your spouse?" On one occasion the potential employee asked what that had to do with his job. The millionaire replied, "If you cannot be faithful to the most important commitment in your life, why would I think you would be faithful and committed to me and my company?" ... Character?

"Leaders with honest self-awareness of their character typically know their limitations and strengths, and exhibit a sense of humor about themselves. They exhibit a gracefulness in learning where they need to improve and welcome constructive criticism and feedback." (Goleman et al., 2004, p. 254)

Every leader in every profession and in our personal lives has past performances in which, if we had the opportunity, would do it differently. In an honest and fair self-evaluation of your character it is best to remember this: it is not just about the fall, more importantly it is about how you get up.

WHO AM I

> *Be yourself; everyone else is already taken.*
> OSCAR WILDE

Strength and Weaknesses (Self-awareness is vital in your assessment of *Who Am I*.)
You Decide:

- What do you honestly believe is your <u>greatest strength</u> in your personal life? _____
- What do you believe you can do to <u>improve and sustain</u> this greatest strength? _____
- What do you honestly believe is a <u>weakness</u> in your personal life? _____
- What do you believe you can do to <u>improve</u> upon this weakness? _____

An Important Beneficial Exercise for YOU: *Also a highly effective exercise for Your Team Members: Your team members will eagerly participate in this role model exercise. Team member comments must be given confidentiality or anonymity.

ROLE MODEL EXERCISE

- Name a person you greatly admire and respect: your role model. This individual may be living or deceased. _____
- Why have you chosen this individual? _____ _____

Your Role Model is going to spend the next six months with you. At the conclusion of the six months there is going to be a banquet with all your friends, associates, relatives, and team members. The keynote speaker is your role model, the subject YOU. Your character, which your role model has observed over the past six months, will be the focus.

- What would you want your role model to say about your character? (Be specific) _____
_____
_____

Present this same exercise to your team members. You will discover they enjoy and take this scenario to heart. It is very effective for their personal discovery of *Who Am I*. Individuals who strive to discover their character will find boundaries, purpose, and direction in leadership and in personal life.

Chapter 4 Goals suggest four team meetings for implementing your mental skills program. In one of these meetings you deem appropriate, inserting the Role Model Exercise will produce numerous benefits for you, your team member, and your team culture.

DEVELOPING YOUR PHILOSOPHIES

Life and Profession

> *The beginning of Philosophy is to know the condition of one's own mind.*
> GREEK MYTHOLOGY

Philosophy Defined: All that is acceptable within a personal boundary, which may include direction and purpose.

Your written philosophy of life will be your guide, your boundaries, and your origin for all you think and act upon. Your philosophy will give you direction in all decision making. It will encompass your moral character developed from your life experiences.

HOW TO DEVELOP YOUR PHILOSOPHY OF LIFE

Having had an enjoyable and successful 38 year coaching/teaching career, I enrolled as a 58-year-old post graduate student at the University of New Mexico in Albuquerque. Our Philosophy instructor, Dr. Hemming Atterbom, gave us an assignment: Develop your philosophy of life. The professor gave some beneficial suggestions, which I pass on to you, that might be helpful in your pursuit.

Go to an environment which is very relaxing and comforting to you, get a Tacate' (Mexican beer), and start to write.

FOLLOWING THE INSTRUCTIONS

Leaving Albuquerque with my laptop and driving to the southern part of New Mexico, I sought a relaxing and peaceful place in the Land of Enchantment. Deciding to splurge, I booked into an excellent motel, Motel 5 ½.

Since I was not a beer drinker, but did enjoy a glass of wine now and then, I stopped at a quick shop and purchased a fine bottle of white wine. This continued spending spree set me back $4.96. Back at the luxurious room in the motel, I sat down at my computer and started to write. Much to my surprise, out of nowhere the words started coming to me. I wrote my personal philosophy of life. My biggest disappointment in this process was my philosophy came so quickly I didn't have a chance to unscrew the aluminum cap on that fine wine, that is at that particular time.

In our next philosophy session back at the University, our professor read each student's philosophy. Having finished one student informed the professor, "You haven't read Sam's." The professor made no response. Then another student responded in like manner. Finally, the professor looked up and said, "You don't understand; I can't read Sam's." My professor had some life's struggles of his own and was fearful of becoming emotional. To this day I believe my professor, whom I admired greatly, had forgotten to dream in his own life.

FAILURE TO DREAM

West Point Sport Psychologist

In a conversation with West Point's head sport psychologist, Dr. Nate Zinsser made the following comment: "I am afraid we are witnessing a culture where many do not know it is okay to dream. Some because they do not know it is okay to dream and some because *fear of failure* in achieving their dreams. I tell the cadets here at West Point to reach for the stars and if you happen to fall short and land on the moon, you still have made a great accomplishment."

In researching and studying great leaders and performers, they all share three commonalities; one of which *they do not fear failure*. Great leaders and performers in all professions and in life have chosen, regardless of circumstances, to continue to dream and pursue their dream.

Your life's philosophy is given birth from all your life experiences. Philosophy of life is developed from one's character interwoven with life's experiences. *Philosophy of Life* comes in many different shapes and sizes. There will be as many different philosophies of life as the number of individuals who read *Mental Skills Leadership*.

The *philosophy of life* I developed while attending the University of New Mexico I still use to this day. This philosophy is presented solely as an example for your thought process in developing your personal written *philosophy of life*.

"A personal relationship with my Creator that permits me to dream and gives me the strength to chase my dreams each new day"

What does this *Philosophy of Life* encompass?

- Purpose: To Dream
- Direction: Forward each new day
- Boundary: Limited strength (implied)

Little did I know at that time in the very near future I would be faced with a one of life's personal challenges far greater than myself. My *philosophy of life* would be my guide through the toughest trials of life. You also have or will experience such monumental challenges; NO ONE ESCAPES this circumstance.

As you develop and assess your personal *philosophy of life*, it should consist of your beliefs your character beckons from the depth of your soul. It is critical to develop a written *philosophy of life*. Having done so, your life and leadership will be more consistently sustainable, successful, and enjoyable.

Caution: There are times in life when one does everything right and desired results are not achieved. That is not failure; that is life. On such occasions your character, your *philosophy of life*, and your ability to refocus will be challenged. Life is not about winning every battle; it is about winning the war.

When things don't go as planned, science of human behavior discovers we have two cognitive choices: Choose to believe you have been persecuted and feel sorry for yourself as long as possible or refocus and make your disappointment a learning process for future success. (*How to Refocus* Chapter 7)

Great leadership and great performers are exactly the same, no difference. Leadership is about performance. Life itself is a performance. The following is an example of refocusing.

PERFECT PITCHING PERFORMANCE AND LOST

Harvey Haddix

One of the greatest performances in baseball pitching history

What do you think: After reading the following story about Harvey Haddix's perfect game, you will be given the opportunity to give your opinion on the following issues:

Harvey Haddix Character
Hank Aaron's Character
Major League Baseball Character
Harvey Haddix ability to Refocus

Question: Do you see any of yourself in the following scenario?

PERFECT PERFORMANCE AND LOST

Harvey Haddix will always be remembered for pitching a perfect game for 12 innings. (At that time Major League Baseball ruled a complete game was 9 innings.) Going into the 13th inning of a game against the Milwaukee Braves on May 26, 1959, Haddix retired 36 consecutive batters in 12 innings relying on two pitches: fastball and slider. However, Pittsburgh also hadn't scored; as the opposing pitcher of the Braves, Lou Burdette, was also pitching a shutout. A fielding error ended his perfect game in the bottom of the 13th, with the leadoff batter for Milwaukee, Felix Mantilla, being safe at first base. Mantilla later advanced to second on a sacrifice bunt, which was followed by an intentional walk to Hank Aaron. Joe Adcock then hit an apparent home run, ending the no-hitter and the game. In all of this confusion Hank Aaron left the base paths and was passed up by Adcock for the second out and the Braves won 2-0. Eventually the umpires changed the hit from a home run to a double, a ruling from the National League. Only Mantilla's run counted, making the final score Milwaukee 1, Pittsburg 0. Haddix's 12 and 2/3 innings, one hit complete game against the team that had just represented the NL in the previous two World Series, is considered by many to be the best pitching performance in major league history.

In 1991, Major League Baseball (MLB) changed the definition of a no-hitter to "a game in which a pitcher or pitchers complete a game of nine innings or more without allowing a hit". Haddix's game was taken off the list of perfect games.

Milwaukee admitted that the Braves pitchers had been stealing the signs from the Pittsburgh catcher, who had exposed his hand signals due to a high crouch. From their bullpen, the Braves pitchers gave their teammates the signal for a <u>fastball</u> or <u>curve ball</u>. Harvey Haddix only used these two pitches during this game. Despite this circumstance the excellent Milwaukee hitters managed just the one hit. All but one Milwaukee hitter, Hank Aaron, took the stolen signals.

The following baseball season Haddix's Pittsburg Pirates won the World Series. He was the winning pitcher of game seven. Harvey

Haddix continued his career with great success until his retirement in 1963. Refocus?

Hummmm ... All but one Milwaukee hitter, Hank Aaron, took the stolen signals. How would you have reacted with the stolen signals if you were on the Milwaukee team? Reading about Harvey Haddix perfect game he lost, what is your opinion about how he refocused the remainder of his career?

After MLB changed the rules that took Haddix's perfect game out of the record book, his comment was "That's alright; I know what I did." Character: Good or Bad? Hank Aaron refused to take the stolen signals. Character: Good or Bad?

Major League Baseball Character

MLB removed Harvey Haddix's perfect game from the record book. Today, however, has MLB turned its head the other way when illegal drugs are used by players and then keeping records on the books achieved with illegal drug use? MLB Character: Good or Bad?

Harvey Haddix's comment, "That's ok; I know what I did", reminds me of the highly talented and successful quarterback of the Dallas Cowboys, Troy Aikman. While living in Dallas, Texas, Troy Aikman's new home burned along with many of his trophies and awards. Troy's comment was, "They can burn my trophies, but they can't take away my memories." Both of these highly respected gentlemen displayed great character in leadership throughout their careers.

Maintaining the balance *What Am I* with *Who Am I* allows an individual to maintain good character. Failure to keep this critical balance enhances the potential loss of your good character. Self-evaluation of your character is paramount in developing your *Philosophy of Life*.

Direction, Boundaries, Purpose

Your *Philosophy of Life* should give you Direction, Boundaries, and Purpose, regardless of the positive or negative circumstances life imposes on each of us. *Philosophy of Life* provides a guideline, a strategy for moving forward.

- Direction–moving forward
- Boundaries—knowing what you can and cannot control
- Purpose–your life's vision

Reverse the Roles: (You are now a team member) How secure would you feel if you depended on someone else for direction, boundaries, and purpose, but none of these criteria (needs) were met? In this scenario someone else is the leader (your pilot) and you are a team member (passenger).

Scenario:

How secure and safe would you feel if the next time you traveled on an airplane your pilot didn't file a flight plan? Obliviously, some have failed to do since they landed at the wrong airport.

As your plane is taking off your pilot announces over the speaker system, "This is your pilot speaking. I know how to take off, fly, and land this plane; but since I didn't file a flight plan I really don't know where I am going to go." How is this pilot's approach (no flight plan) any different than your leadership without a personal written *philosophy of life* (flight plan)? No direction, no boundaries, no purpose. If you have no flight plan, so too will your team members. Your team members expect you to have a flight plan.

Remember how you would feel about this circumstance. This is exactly how your team members will feel if you and your leadership do not provide direction, boundaries, and purpose. Every action, every policy, and every thought in life and profession will originate from your philosophy of life. Your written *philosophy of life* and your written philosophy of profession are significant for all you seek to accomplish. Your team members expect a strategy for pursuing successful performance.

Life's Priorities

From the list of examples choose in order three top priorities in your personal life; not as they should be, but as they are presently:

Examples: Profession, Family, Personal Relationships, Country, Your Creator, Other

1._____ 2._____
3._____

List how you think your priorities really should be in your life

1._____ 2._____
3._____

*Should your answers <u>not</u> be identical you now have discovered an aspect to improve. Should your answers be identical you have reaffirmed strength.

DEVELOPING YOUR PERSONAL PHILOSOPHY OF LIFE

Know Thy Self
SOCRATES

ESSENTIAL FOR SUCCESSFUL AND SUSTAINABLE LEADERSHIP

Philosophy Requirement: An honest fair self-evaluation of your mind and heart

- Find a relaxed peaceful environment
- Eliminate all distractions
- Focus on *Who Am I* (self-awareness and evaluation of your character)
- Be brutally honest, however make sure you are fair to yourself

Your written *Philosophy* will be unique and, as well as your leadership. Your *philosophy of life* is deeply embedded in your character and from the totality of your life experiences: (*Who Am I*)

YOUR PHILOSOPHY OF LIFE

Develop *Your Philosophy of Life*: _____
_____
_____
_____
_____

One of the important benefits in developing *Your Philosophy of Life*: IT BECOMES an enabler for developing your ability to refocus. Having developed your *Philosophy of Life*, you are now prepared to develop *Your Philosophy of Profession*.

DEVELOPING YOUR PHILOSOPHY OF PROFESSION

I've always coached the way I wanted to be coached.
TONY DUNGY, former NFL coach

Philosophy of Profession: Perception of your purpose in your profession

Since processing a careful self-evaluation in seeking *Who Am I*, you are now capable of developing your *Philosophy of Profession*: your purpose. Coach Dungy's philosophy is an excellent example for your consideration in constructing your *philosophy of profession*. This great leader's philosophy (purpose) was *to lead the way he wanted to be led*. (Lynch & Chungliang, 2006)

Assessing the following generic philosophies, are you satisfied with the example or would you improve the philosophy? Do you see any similarities or differences of philosophies between these professions?

Generic Examples

Politics- To serve others; Business- Create profit while maintaining good character; Military- Personal privilege and duty to defend freedom; Education- Teach students to sing in perfect harmony; Sport- Successful performance with character building; Religion- I am a servant of those I serve

Applied Philosophy of Profession

Former Army Chief of Staff Gordon Sullivan and Michael Harper wrote an excellent book on leadership, *Hope Is Not A Method*. (1996) In this writing they tell the true story of a happening at Appomattox during the American Civil War.

Appomattox

We were ragged and had no shoes. The banners our Army had borne to the heights of Gettysburg were bloody and in shreds.... We were only the shadow of an army, the ghost of an army, and as we marched in tattered, hungry columns between those magnificent straight lines of well-fed, faultlessly armed and perfectly equipped solders, most of us wished, as our great chief did, that we might have numbered with the fallen in the last battle...

Suddenly I heard a sharp order down that blue line, and on that instant I saw the whole brigade present arms to us——to us, the survivors of the Army of Northern Virginia. It was a Maine brigade, comrades, and I confess to you that......I never hear the name of that state but that I feel a certain swelling pride as I reflect that there was an army good enough to deserve that salute——and another magnanimous enough to give it.

You Decide

What was this great military leader's *Philosophy of Profession*? **Winning with Character**? What if you are ever in such positions in your leadership: What will you do? Is there a value here for leadership? Giving serious thought in developing or reviewing your *Philosophy of Profession* will be a strong indicator how you will respond in like circumstance.

PERCEPTION OF YOUR PURPOSE IN YOUR PROFESSION OF LEADERSHIP

YOUR PHILOSOPHY OF PROFESSION

(Your Thoughts) _____

_____

_____

Philosophy of Life, Philosophy of Profession a solid foundation for all you wish to achieve

*Becoming a Great Leader is not the hardest part; the hardest part is Sustaining Your Great Leadership without destroying the rest of Your Life

SECURE LEADERS VS INSECURE LEADERS

The person who is not afraid of oneself is more effective than the person whom everyone fears.
T. HUGHES

Two Toxic Tenets of Leadership

- **Hubristic Leader:** An individual drunken with power.

This type individual cares about only one thing: Self. Performance goals are a solution for eliminating insecurity. However, addressing a solution for hubris beckons my recall of a conversation with a military general. In this conversation the General made the following comment,

"In war we can defeat everything except ignorance; therefore our only option is to destroy it." Eliminating hubris presents us with only one option, brain transplant.

- **Insecure Leaders:** Fundamentally, insecurity is fear.

This type requires having their ego puffed up by as many team members which can be convinced, coerced, or threatened to do so. Insecure leaders cause damage beyond human measurement. Whatever good qualities there may be in a team, the insecurity of an individual in a leadership position will eventually erode and destroy each one of those positive qualities. I would elaborate in more detail upon the damages caused by insecure individuals in leadership positions, but there is no doubt the rise of my blood pressure would reach the outer limits of our solar system. If team members have to serve under insecure leaders, their blood pressure would eventually rise to infinity and beyond. Insecure individuals in leadership position cause destruction far beyond human explanation.

- **Secure Leaders:** Self-confident leaders keep their ego in check for the benefit of improved performance of each team member. Secure leaders create norms and develop teamwork, harmony, and unit cohesion. Doing this will release a powerful force in the pursuit of your visionary goal. (Goleman et al., 2004)

Solution for minimizing or eliminating one's own insecurity in leadership: Develop a written goals program with the purpose of increased self-confidence. Self-confident leaders are competent leaders. Building self-confidence decreases or eliminates insecurity.

Proper goal setting is fail proof. It is possible you may lose a battle or two, such as Harvey Haddix, but you will win the war. Secure leaders do not need their ego puffed up, they know who they are. Insecure individuals in leadership positions need to have their ego stroked, which results in tearing down or holding down others in fear of team members receiving credit for their own success. Insecure leaders permeate a toxic atmosphere throughout the organizational structure.

YOU ARE THE JUDGE

(Secure or Insecure)

Which of the following statements apply to a Secure or an Insecure individual in leadership?

1. A leader promotes their agenda at any cost——

    SECURE   INSECURE

2. A leader that understands you treat team members with respect, compassion and firmness——

    SECURE   INSECURE

3. A leader that promotes their leadership based on fear——

    SECURE   INSECURE

4. A leader who understands and implements the art of coaching——

    SECURE   INSECURE

5. A leader who demands all success be credited to themselves——

    SECURE   INSECURE

6. A leader who promotes ownership for each team member——

    SECURE   INSECURE

7. A leader who desires B+ team members——

    SECURE   INSECURE

8. A leader who desires and develops A+ team members——

SECURE   INSECURE

9. A leader with tight fisted control (control freak)——

SECURE   INSECURE

10. A leader with boundaries of fairness and expectations——

SECURE   INSECURE

11. A leader who perceives team members as liabilities——

SECURE   INSECURE

12. A leader who values team members——

SECURE   INSECURE

13. A leader who wants the credit——

SECURE   INSECURE

14. A leader who does not view disagreement as insubordination——

SECURE   INSECURE

15. A leader who views disagreement as insubordination——

SECURE   INSECURE

Answers: Odd numbers insecure leaders:
Even numbers secure leaders

AN INDIVIDUAL WITH A VISIONARY GOAL THAT HAS THE ABILITY TO MOTIVATE TEAM MEMBERS TOWARDS ACCOMPLISHING YOUR VISIONARY GOAL

CHAPTER 1 AN INDIVIDUAL: YOU

SUMMARY

Character Determines Sustainability of Success in Profession and Personal Life

The importance of your personal evaluations concerning the contents of this chapter is immeasurable in your pursuit of the highest level of sustainable leadership. A systematic evaluation of your character is invaluable in your profession and in your personal life. Of all the evaluations we make in life, none is more critical and important than those we make about ourselves. We are exactly as we perceive our self to be.

CHAPTER 1 AN INDIVIDUAL (YOU): MAJOR TOPIC POINTS

WHO AM I

Self-Evaluation of Your Critical Balance:

- What Am I vs. Who Am I (A Critical Balance necessary for Leadership and Life)
- Inevitable Self-Destruction (Imbalance of Character with Profession)

What Am I

- Your Profession
- Stolen Identity (When the *What Am I* consumes the *Who Am I*)

Who Am I

- Self-evaluation of Your Character (Strengths and Weaknesses)

DEVELOPING YOUR PHILOSOPHIES:

Your Philosophy of Life

- Direction, Boundary, Purpose
- Developing Your Philosophy of Life

Your Philosophy of Profession

- Perception of Your Purpose in Your Profession of Leadership

- Developing Your Philosophy of Profession

Secure Leaders vs. Insecure Leaders

- Toxic Tenets Hubris-Insecurity

CHAPTER 1 GLOSSARY: AN INDIVIDUAL (YOU)

Hubris: Drunken with power

Insecure Leader: Fundamentally insecurity is fear. Leaders requiring having their ego puffed up by as many team members that can be convinced, coerced or threatened to do so

Philosophy of Life: All that is acceptable within a personal boundary (may include purpose and direction)

Philosophy of Profession: Perception of your purpose in your profession

Secure Leader: Self-confident leaders keep their ego in check for the benefit of improved performance by each team member

Stolen Identity: When the Who Am I is consumed by the What Am I

What Am I: Your Profession

Who Am I: Your Character

Your Character is the Alpha and Omega of Your Leadership and Your Personal Life

Everybody ages. Everybody dies. There is no turning back the clock. So the question in life becomes: What are you going to do while you're here?
GOLDIE HAWN International Entertainer

CHAPTER 2:

VISIONARY GOAL: YOURS

◆ ◆ ◆

INTRODUCTION

AN **INDIVIDUAL** WITH A **VISIONARY GOAL** THAT HAS THE ABILITY TO **MOTIVATE TEAM MEMBERS** TOWARDS ACCOMPLISHING **YOUR VISIONARY GOAL**

Pick and choose what you fall on your sword for.
WEST POINT MILITARY ACADEMY

Following the linear sequence of bold words in the definition of leadership: Chapter 1 discussed the **Individual** (You); Chapter 2 presents developing **Visionary Goal** (Yours)

The pursuit of your vision requires sacrifices. This is the essence of the Colonel's statement: "Pick and choose what you fall on your sword for." Your visionary goal is the foundation for all you and your team members attempt to achieve in your organizational structure. Pursuing successful leadership to its highest level is not the greatest challenge; the greatest challenge is sustaining your level of successful leadership.

Even though you are on the right track you will get run over if you just sit there.
WILL ROGERS

CHAPTER 2 YOUR VISIONARY GOAL: MAJOR TOPICS

YOUR VISIONARY GOAL

- **Empowerment** (Importance of Empowering Team Members)
- **Leadership Goals (2)** (Your Vision; Your Strategy)
- **Five Leadership Mistakes** (with solutions)
- **Chapter 2 Glossary:** Empowerment, Performance Goal, Team Culture, Visionary Goal

VISIONARY GOAL: YOURS

AN **INDIVIDUAL** WITH A **VISIONARY GOAL** THAT HAS THE ABILITY TO **MOTIVATE TEAM MEMBERS** TOWARDS ACCOMPLISHING **YOUR VISIONARY GOAL**

Visionary Goal Defined: A vision and a sense of the future. It is an imagined possibility, stretching beyond today's capability, providing an intellectual bridge from today to tomorrow, and forming a basis for looking ahead, not for affirming the past or the status quo. (Sullivan & Harper, 1996)

Great leaders are always seeking new techniques, theories, and visions for improvement. Not only will your **visionary goal** put you on the right track, it will empower you and each of your team members to move forward in successful pursuit of **accomplishing your visionary goal**.

With a strong sense of your character and your purpose of profession you are now capable of developing a written clearly defined *visionary goal*. (Lynch & Chungliang, 2006) Great leaders create a written *visionary goal* and clearly articulate this vision to team members with passionate ownership and consistent pursuit of its success. (Sullivan & Harper, 1996) When you craft a *visionary goal* with positives and heartfelt passion you will inspire a team vision interwoven throughout your organizational structure, enabling team members' understanding their assigned performance is meaningful.

YOUR VISIONARY GOAL

Requires two technical necessities

- Written / Clearly Defined (simplicity)
- Effectively Communicated Vision to Team Members

Your leadership must give clarity to *your visionary goal*. If your team members do not understand the whole *visionary goal*, they won't understand their piece of your *visionary goal*.

It is vital to effectively communicate the *visionary goal* to each team member. Team members who do not understand *your visionary goal* will not have purpose, ownership, or empowerment in successful pursuit of *your vision*. Failure to effectively communicate *your visionary goal* will greatly handicap its success.

Your Visionary Goal: There are more conflicts and issues within a team organization than between organizations. Your written *visionary goal* will dramatically minimize these conflicts and issues. The same is true with family structure. There are more issues within a family structure than between family structures.

Developing a written *visionary goal* with clarity and effectively communicating to team members, empowers and unleashes the talents of each team member to achieve the future. In the final analysis everything comes back to each team member and how they work together. (Teamwork-Harmony-Unit Cohesion) Structuring simplicity in the *visionary goal* is a lesson we can utilize from the elite missions of the military.

Leaders become great not because of their power, but because of their ability to empower others. (Maxwell, 2011)

EMPOWERMENT

Your *visionary goal* enables you to develop positional boundaries and positional expectations throughout your organization; in doing so you have established a solid structural foundation for empowering each team member.

Empowerment Defined: Leadership providing team member with necessary skills; physical, technical, and mental, for successful performance of positional expectations within their positional boundary.

> *If I have achieved greatness it is because I have stood on the shoulders of Giants.*
> STAN MUSIAL, Baseball Hall of Fame

Musial attributes managers (leaders) empowering him, which enabled a highly successful career and membership to the Baseball Hall of Fame. Stan Musial was a man of Great Character.

Elevating leadership to its highest sustainable level necessitates inclusion of empowering team members. Empowering team members to act within positional boundaries with acceptable behavior as defined by your values, your vision (Your Visionary Goal), your strategy (Your Performance Goal), will elevate your leadership to its highest sustainable level.

Clearly communicating your Visionary Goal (vision) and your Performance Goal (strategy) empowers team members to pursue your vision with greater self-confidence, purpose, and motivation. Each team member needs to know your leadership expectations within their positional boundary. Establishing team member positional expectations is the beginning of empowerment. Empowerment is about team member responsibilities within their positional boundaries in which the team member has controllability over failure or success. "Treating team members with dignity and respect is the keystone of effective team building, future leader development, and empowerment." (Sullivan & Harper, 1996, p. 232)

Thought Provoking Exercise

The following space is provided for you to describe <u>one</u> team member positional boundary and your expectations within this positional boundary. Description of Team Member **Positional Boundary**

Description of Team Member **Positional Expectations**

In the previous exercise you described a position and its expectations. If you were to ask one of your team members, in this same position, what they believe their positional boundary and their positional expectations are, do you think you and your team member will be on the same page? Once you empower your team member with positional boundary and positional expectations, it becomes time for that team member to realize their responsibility for failure or success of performance. (Corey, 2011)

With all things being equal; know thy self (*Who Am I*) and you will be successful 50% of the time, know your opponent and you will be successful 100% of the time. (Lynch & Chungkuang, 2006, p. 39)

Translated: Know thy self (*Who Am I*) and you will be successful 50% of the time, know your team members (*Who Am I*) and you will be successful 100% of the time.

Clarifying team member positional expectations empowers each team member in pursuit of accomplishing your visionary goal. This clarity provided by your leadership enables team members to set appropriate destination and performance goals. (Chapter 4: Goals)

CHECKLIST FOR EMPOWERMENT

What are your Team Members **Positions**?

√ Have You Clearly Defined each **Positional Boundary**?

- √ Have You Clearly Defined your Team Members **Positional Expectations**?
- √ Does each of your Team Members Know and Understand their **Positional Expectations**?
- √ Have you Empowered each of your Team Members with Skills necessary for Successful Achievement within **Positional Boundaries**? (Physical, Technical, Mental)

Every team member has special talents. It is *your* responsibility as their leader to create a team culture in which each team member can discover and apply their special talents within their positional boundaries. Empowerment within positional boundaries is a necessity providing numerous important benefits to your team culture. It is important your leadership discovers the good in each team member. To paraphrase a great American President: If one looks for the bad in someone, one will surely find it. If one looks for the good in someone, one will surely find it. It is your leadership's role and duty to look for the good in each team member. Not doing so is counterproductive in successful pursuit of *your leadership's visionary goal.*

THE WRECKING CREW

Looking for the good in team members

Early in my career I coached a high school basketball team, in which the second five were affectionately named The Wrecking Crew. This team accomplished a season record of 24 wins and 2 defeats. The two defeats were our first two games.

The first seven team members of the 12 man squad had excellent basketball abilities. The final five team members were basketball-ability-challenged. However, they had wonderful attitudes. These final five team members were important and vital in accomplishing our visionary goal.

Our performance goal (strategy) was to work as hard in practice as we did in a game.

As leader of The Wrecking Crew my *Visionary Goal* was: Develop a team culture that empowers and motivates each team member to perform their best throughout the season, regardless of circumstances. Without The Wrecking Crew how would we be able to have practice scrimmages? How could we empower them and bring them into a successful team culture?

In practice scrimmage they made the first five prepare far beyond our preparation for the next opponent. On defense the second five seldom went where they were supposed to go; on offense, pot-luck, shot-gun performance. We didn't know where they were going to go. In some instances they didn't know where they were going to go. However, wherever they went they did so with passion.

These basketball-ability-challenged team members gave themselves the following appropriate name: The Wrecking Crew. If The Wrecking Crew could discombobulate our offense and defense, what would they do to our opponents? An effective strategy (performance goal) was needed enabling the *Visionary Goal* to be successful with inclusion of the Wrecking Crew's passion.

We decided to develop a more detailed strategy (Performance Goal) to empower The Wrecking Crew. We'd start The Wrecking Crew; allow them to play the first quarter of each game as long they keep the score close. This decision included input and approval from the remaining team members. The Wrecking Crew accomplished this goal, playing the entire first quarter of each game until the final game of the season.

This final game was our third championship game of the season. At the four minute mark in the first quarter our opponent began to open up a margin, requiring the first five to prematurely enter the game. As the game progressed, with 2:25 left in the third quarter our opponent opened up a 26 point lead. Our opponent's first five were 6-6, 6-6, 6-5, 6-4, and 6-2; each of these opposing players could slam dunk the ball. Their school enrollment was 2,300 and our school enrollment 305.

Our team members were 5-11, 5-11, 5-10, 5-9, and point guard 5-4, none of which could slam dunk the ball. With one second remaining in the game, we scored and won by one point. None of these accomplishments could have been achieved had it not been for finding the good in The Wrecking Crew.

What did playing these team members in the first quarter of each game accomplish? What were the positive results of finding the good in these team members (The Wrecking Crew)? Are there such circumstances within your professional leadership?

EMPOWERMENT: A VITAL PART OF THE TEAM CULTURE (OWNERSHIP, PURPOSE, MOTIVATION)

The first seven players were on the same page knowing how important The Wrecking Crew was to the team culture and the team success in accomplishing the *visionary goal*. In other words our *visionary goal* was effectively and clearly articulated for each team member.

Regardless of profession you most likely have team members analogous to the members of The Wrecking Crew. Finding the good in each team member and placing this ability in the proper organizational position is vital to the success of *Your Visionary Goal*. Each team member has special talent. It is your leadership's responsibility to assist each team member to discover this talent and place their talent in the most effective team position. Having done so, you have empowered your team members.

The science and art of human behavior is important in leadership. Great leaders perceive this as an exciting challenge. How to discover each team member's needs and special talent will be discussed in the mental skill of setting goals, Chapter 4.

YOUR LEADERSHIP GOALS

(1) Visionary (2) Performance

As leader you will develop two goals

1. **Leadership Visionary Goal:** (Your <u>Vision</u> of Future Accomplishment)
   Leadership Visionary Goal: a long term goal crafted for what you and your team members expect to accomplish in a predetermined time frame. When you the leader pursue the *visionary goal* with passion, your passion will become contagious within your team culture.

2. **Leadership Performance Goal:** (<u>Strategy</u> to Achieve Your Visionary Goal)
   Leadership Performance Goal describes your strategy for pursing successful accomplishment of your *visionary goal*. *Leadership Performance Goal* describes how you intend to transcend your *visionary goal* into reality with team member performance.

*Your *Leadership Visionary Goal* is vital, so too is your strategy *Leadership Performance Goal*

Strategy Defined: A cognitive, tactical evaluation of where you are and designing a process to achieve the future: your vision. Team members expect their leader to have a vision and a strategy for them to pursue excellence.

From the following list of Professions, observe three professions:
- First: Life
- Second: Your Profession
- Third: Another of your choosing from: (Education, Business, Military, Politics, Sport)

Give your thoughts to the generic *Leadership Visionary Goal* and the *Leadership Performance Goal*. Do you see any need for improvement in the following generic goals?

Life Generic Example:

Life Visionary Goal: Sincerely be conscious of becoming more understanding and compassionate towards those I love.

Life Performance Goal: Focus on the positive blessings I receive from my loved ones.

Your Thoughts on the above Life Goals:

In your personal life, what is your *visionary goal*?

What is your *performance goal* in your personal life; strategy to accomplish your Life Visionary Goal?

Business Generic Example:

Business Visionary Goal: Increase our yearly production 8-12%.

Business Performance Goal: Evaluate and reposition team members in team positions of best fit without additional costs.

Your Thoughts on the above Business Goals:

In a business leadership role, what would be your *visionary goal*?

What is your *performance goal*; strategy to accomplish your Business Visionary Goal?

Sport Generic Example:

Sport Visionary Goal: Our season's visionary goal is 21-27 victories.

Sport Performance Goal: Develop a team culture of harmony and motivation by effective team member written performance goals including leadership feedback.

Your Thoughts on the above Sport Goals:

In your sport leadership role, what is your *visionary goal?*

___

What is your *performance goal,* strategy to accomplish your Sport Visionary Goal?

___

Education Generic Example:

Education Visionary Goal: Create an environment for improved teacher-parent relationship throughout the school year.

Education Performance Goal: Each teacher conduct parental classroom visitation with parental feedback 2-5 times yearly.

Your Thoughts on the above Education Goals:

In an educational leadership role, administrator, teacher what is your *visionary goal?*

___

What is your *performance goal,* strategto accomplish your Educational Visionary Goal?

___

Military Generic Example:

Special Note: During the writing of *MSL* elite coalition military forces completed several successful missions in major military theater overseas. On national television the former leader of this team was asked the following question: How did your former team pull off such successful missions without any causality? "We always have clarity and simplicity of mission."

Military Visionary Goal: Our mission can best be accomplished with unit cohesion (teamwork, harmony).

Military Performance Goal: Give clarity of visionary goal, and positional expectations to empower each team member.

Your Thoughts on the above Military Goals:

In a military leadership role, what is your *visionary goal?*

_____

What is your *performance goal*; strategy to accomplish your Military Visionary Goal?

_____

Politics Generic Example:

Political Visionary Goal: Focus on the needs of constituents regardless of political affiliation.
Political Performance Goal: Address constituents need then develop and pass non-partisan bills.

Your Thoughts on the above Political Goals:

In a political leadership role, what is your *visionary goal?*

_____

What is your *performance goal*; strategy to accomplish your Political Visionary Goal?

_____

Bobby Knight

Love him or hate him, Bobby Knight's Leadership was one of the greatest *strategists* in college sports. Give him an extra day to plan strategy (performance goal) for the next opponent and his team could compete with anyone, anywhere, anytime. Ability to develop the proper strategy (*your leadership performance goal*) is crucial to the achievement of your *visionary goal*.

Great strategists successfully address every detail. In a 2011 SEC Championship game a television commentator made the follow observation about the leadership of one of the head coaches, "No other leader pays as much attention to details in preparing his team's performance."

BALD EAGLE'S STRATEGY

After a day's flight the bald eagle will preen every feather. This deliberate, self-disciplined, detailed effort is the strategy for the next day's performance. The Bald Eagle's strategy is synonymous with leaders who develop successful detailed strategy for a desired performance.

Those who have developed successful, sustainable leadership in profession have numerous talents and benefits to offer in any profession. We have often wondered why more great leaders have not entered politics or have been chosen by politicians to serve on their team.

FIVE COMMON LEADERSHIP MISTAKES WITH SOLUTIONS

1. **Tight Fisted Control** (Control Freak) Tight Fisted Control is micro management on steroids. Inexperienced individuals in a leadership role often implement extremely tight control which may be referred to as "control freak".

This type of leadership is also implemented by an individual who is insecure, lacks adequate self-confidence or is drunk with power-hubris. Control Freak damages team member's ownership, purpose, self-confidence, empowerment, team culture, and team cohesion.

> **Solution:** Develop a personal goals program for the purpose of improving self-confidence. (Chapter 4 Goals)

2. **Unclear or Lack of Visionary Goal** Visionary goals that are unclear and that do not provide definite boundaries or clarity of purpose is a breeding ground for multiple issues. It is also possible the visionary goal is proper, but your leadership fails to effectively communicate your vision to team members with clarity.

   > **Solution:** A written *visionary goal* effectively communicated is essential for successful, sustainable leadership. Creating a duplicate picture of your visionary goal in the mind of each team member is a necessity. In every instance a duplicate picture may be verified by leadership feedback. (Chapter 4 Goals details proper feedback)

3. **Undeveloped Positional Boundaries** It is imperative each Team Member understands the positional boundary of their assignment. Unclear positional boundaries do not permit team member control within their position, which creates numerous issues. Unclear positional boundaries are confusing and harmful to setting proper goals for team members who ultimately pursue the success of your visionary goal … Chaos!

   > **Solution:** Written clearly defined boundaries for each position. A boundary that encapsulates position expectations

4. **Failure to Build a Team Culture**

Culture Defined: Shared Norms and Values

In failing to build a team culture the following important benefits will not occur: Team Work, Harmony, Unit Cohesion, and Empowerment.

"Norms and deviance are socially important because they help define and regulate the boundaries of a team. Shared visible allegiance to a culture's norms heightens awareness of boundaries by violating them. In so doing, they reinforce the sense of teamwork, harmony, unit cohesion, empowerment, and belonging for those team members who conform. Culture is a part of the social environment in which team members views take shape." (Johnson, 2000, p. 246)

> **Solution:** Successful team culture requires each team member be on the same page in pursuit of your *visionary goal*. Proper *team member goal setting* is a tremendous building block for each team member being on the same page. (Chapter 4 Goals)

Team Culture: Great Leadership begins with good character. "Emphasizing good character values signals what will not change, providing an anchor for team members drifting in a sea of uncertainty and a strategic context for decisions and actions that will grow the team." (Sullivan & Harper, 1996, p. 64) Without developing a team culture your leadership *visionary goal* is severely compromised.

5. Attempting to Please Everyone All The Time

OLD MAN...LITTLE BOY...DONKEY

There was an old man, a boy, and a donkey. They were going to town and it was decided that the boy should ride. As they went along they passed some people who exclaimed that it was a shame for the boy to ride and the old man to walk.

The man and the boy decided that maybe the critics were right so they changed positions. Later they passed more people who then exclaimed that it was a real shame for the man to make such a small boy walk.

The man and the boy decided maybe the critics were right so they decided that they both should ride. They soon passed other people who exclaimed that it was a shame to put such a load on a poor little animal.

The old man and boy decided that maybe the critics were right so they decided to carry the donkey. As they crossed a bridge they lost their grip on the animal and the donkey fell into the river and drowned. (Martens, 1997)

Moral to the story: If you try to please everyone all the time you will eventually lose your *rass*.

> **Solution:** Daily review of *Your Visionary and Performance Goal* with strict adherence to their purpose

*****Caution:** It is necessary for these common leadership mistakes to be addressed and eliminated for sustainable success of *Your Leaderships Visionary and Performance Goal*

AN **INDIVIDUAL** WITH A **VISIONARY GOAL** THAT HAS THE ABILITY TO **MOTIVATE TEAM MEMBERS** TOWARDS ACCOMPLISHING
YOUR VISIONARY GOAL

CHAPTER 2 VISIONARY GOAL: YOURS

SUMMARY

Efforts and Courage are not enough without Purpose and Direction.
JOHN F. KENNEDY

Leadership without a clearly articulated Visionary Goal permits an undisciplined environment in which your team members will be left in a fog, with confusion and lack of direction. Three major benefits of Visionary Goal: purpose, direction, and boundary. Effectively communicating your written Visionary Goal to each team member allows team members to be on the same page in successful pursuit of the ultimate; Your Visionary Goal.

CHAPTER 2 VISIONARY GOAL: YOURS
MAJOR TOPIC POINTS

Your Visionary Goal: (uniquely yours) two technical necessities

- Simplicity: Written, Clearly Defined
  Effectively Communicated: You the leader must effectively communicate your visionary goal to enable each team member understanding their piece of your visionary goal

Empowerment: (Importance of Empowering Team Members)

Checklist

What are your Team Member Positions?

Have You Clearly Defined each Position Boundary?

Have You Clearly Defined your Team Members Positional Expectations?

Does each of your Team Members Know and Understand their Positional Expectations?

Have You Empowered each of your Team Members with Skills necessary for Successful Achievement within Positional Boundaries? (Physical, Technical, Mental)

Empowerment Applied

The Wrecking Crew: Finding the good in each team member

Leadership Goals (2)

Visionary Goal (Yours)

- Your Vision
- Long term goal crafted for what you and your team members expect to accomplish in a predetermined time frame

Performance Goal (Yours)

- Your Strategy
- It is your leadership's written strategy for accomplishing your visionary goal

Other Professions: Visionary Goal with compatible Performance Goal

- Life
- Business
- Sport
- Education
- Military
- Politics

Five Common Leadership Mistakes with Solutions:

Tight Fisted Control or Control Freak, Unclear or Lack of a Visionary Goal, Undeveloped Boundaries, Failure to Build a Team Culture, Attempting to please everyone all the time

CHAPTER 2 GLOSSARY: VISIONARY GOAL

Empowerment: Leadership providing team member with necessary physical, technical, mental skills for successful performance within positional boundaries

Performance Goal: Leadership's strategy to achieve the visionary goal

Team Culture: Shared norms and values

Visionary Goal: An imagined possibility, stretching beyond today's capability, providing an intellectual bridge from today to tomorrow

Your Visionary Goal should Transcend Your Team Culture.

As important as *your visionary goal* and *performance goal* are, never forget your success comes from team members. *Your Visionary Goal* gives purpose, direction, and boundary for each team member.

There is a price to be paid to elevate your leadership to its highest sustainable level.

CHAPTER 3:
MOTIVATE TEAM MEMBERS

◆ ◆ ◆

INTRODUCTION

AN **INDIVIDUAL** WITH A **VISIONARY GOAL** THAT HAS THE ABILITY TO **MOTIVATE TEAM MEMBERS** TOWARDS **ACCOMPLISHING YOUR VISIONARY GOAL**

Creating a Motivational Environment

Before an individual can be motivated that individual must perceive they have the ability to be successful

Following the linear sequence of bold words in the definition of leadership Chapter 1 **Individual** (you); Chapter 2 **Visionary Goal** (yours); Chapter 3 **Motivate Team Members**

Chapter 1 presented a step by step method for creating a foundation to enable self-discovery of your true character in achieving the critical balance of *What Am I* with *Who Am I*. Your character permeates throughout the entity of your leadership in profession and in life.

Chapter 2 offered a process enabling you to develop a written, clearly articulated *visionary goal* for your organizational team. *Your visionary goal* provides purpose, direction, and boundaries for your team structure.

Chapter 3 discusses **Motivate Team Members** in developing a motivational environment for your team members' in successful pursuit of your visionary goal. Perceiving future success is a precursor for motivating team members. Presented are the necessary tenants for creating a successful *Motivational Environment*.

CHAPTER 3 MOTIVATE TEAM MEMBERS: MAJOR TOPICS

Science of Motivation: Knowledge

Two Types of Motivation
Two Theories
Criteria for Motivational Environment

Art of Motivation: Application

Effective Communication
Develop Performance Goals Environment
Leadership Skills to Consider
Issues to be Addressed in Your Motivational Environment

Chapter 3 Glossary: Additive Principle, Coaching, *Carpe diem*, Competence, Counseling, Eidetic, Effective Communication, Environment, Extrinsic, Intrinsic, Motivation, Multiplicative Principle, Nepotism, Public Criticism, Revenge, Self-discipline, Subjective Awards, Team

MOTIVATE TEAM MEMBERS

CREATING A MOTIVATIONAL ENVIRONMENT

AN **INDIVIDUAL** WITH A **VISIONARY GOAL** THAT HAS THE ABILITY TO **MOTIVATE TEAM MEMBERS** TOWARDS ACCOMPLISHING **YOUR VISIONARY GOAL**

Motivating Team Members is not only a science but also an art. **Science** is the knowledge and **Art** is the successful application. Topics in this chapter address both science and art for *creating a motivational environment*.

Team: *E Pluribus Unum* "Out of Many, One"

Motivation: The degree of *mental intensity directed towards* the accomplishment of a *goal*.

Environment: A carefully crafted atmosphere providing boundaries from which a team member may pick and choose from a selected menu of acceptable stimuli, designed for specific performance expectation.

Science of Motivation: Knowledge

The accomplishment of motivating team members is greatly enhanced when you develop and provide the correct environment. When you *create a motivational environment* for team members, they will perform naturally without being coerced to do so. Failure to accomplish the primary task of motivating team members in the proper direction ... nothing else within your team structure will work as well as it could or should. (Goleman et al. 2004)
*It is important to remember building self-confidence is a necessary precursor to motivation.

DECREASING MOTIVATION

A beautiful new track and field stadium was built by an Olympic Gold Medalist. We had the opportunity to compete in this wonderful arena since it was in our sports area. What a great extrinsic motivator ... or so we thought.

The first year our men and women's track and field team participated, the sign at the stadium entrance read: NO ALCOHOL BEVERAGES. The second year the sign read: NO ALCOHOL BEVERAGES, NO KNIVES. The third year the sign read: NO ALCOHOL BEVERAGES, NO KNIVES, NO GUNS. The fourth year we didn't go back! *Lesson learned: To motivate team members you do not place mental obstacles in their path; you remove them.

"An environment is only a potential until acted upon by appropriate actions of the team member. It is not a fixed property that inevitably impinges upon team members. For example, your visionary goal will not be acted upon unless team members attend and participate in the motivational environment. Books do not affect people unless they are read. A motivational environment

is a two way street. Leadership provides a motivational environment, team members participate." (Bandura, 1977, p. 195)

It is leadership's responsibility to place team members in a position to succeed. When your leadership develops team members with elevated motivation, they will often outperform others with more talent who lack such quality.

The purpose in developing a *motivational environment* is to enable team members to successfully pursue, with passion and intensity, your visionary goal. Developing motivation in team members is linked to self-confidence. Self-confidence is synonymous with personal competence. **Competence Defined:** Competence is a team member's conviction that a successful outcome can be achieved. (Cox, 2011)

A GAGGLE OF GEESE

Motivational Environment

The importance of harmony, teamwork, and unit cohesion is evidenced with the V- formation of geese. Scientists have implemented technology to discover when geese fly as a team in V-formation they travel 71% further than traveling solo or tandem. (Lynch and Chungliang, 2006) Your *motivated* team flying in formation (teamwork) will travel much further in achieving *your visionary goal*.

Also discovered when a member of the gaggle becomes injured or sick, one of the geese will remain to assist in fulfilling the needs of the injured. When the injured has recovered enough, they rejoin the V-formation. This is but two of many lessons leadership in profession and life may benefit from the animal kingdom.

Two Types of Motivation

1. Intrinsic: Motivation comes from *within* an individual (team member) "Motivation to participate in a performance for its own sake and for no other reason." (Cox, 1985, p. 204)

- Develops a longer duration of motivation compared to extrinsic motivation.
- Is a process not an event
- Develops sustainability
- Enables a mental environment for developing the ability to refocus. The inability to refocus is the number one cause of failure in profession and in life. (Chapter 7 Refocus)

The ultimate goal of *creating a motivational environment* is to develop *intrinsic* motivation towards the achievement of your visionary goal. Chapter 4, Goals identifies one of the products of team member goals as enabling a mental atmosphere for developing *intrinsic* motivation.

2. Extrinsic: Motivation comes from an *outside* source such as external rewards to motivate behavior in performance: money, ribbons, medals, trophies, praise, etc. *Extrinsic* motivation may be an event(s) which can be used to develop intrinsic motivation. Furthermore, *extrinsic* motivation *has a shorter duration* when compared to intrinsic motivation.

Two Theories

Additive Principle Defined*:* The notion *intrinsic* and *extrinsic* motivation are additive to create need achievement.

> *Additive Principle*: According to the Additive Principle, a performer who is low in achievement will participate in an achievement situation if there is sufficient reward or *extrinsic* motivation for doing so. (Cox, 2011) However, recently a great deal of research seems to question the Additive Principle.

Multiplicative Principle Defined: The belief *intrinsic* and *extrinsic* motivations are interactive and not additive.

> *Multiplicative Principle*: Some scientists believe the relationship between *intrinsic* and *extrinsic* motivation is *multiplicative*, not additive. In other words *extrinsic* can either be helpful or harmful to *intrinsic* motivation. (Cox, 2011) *Extrinsic* motivation would have credibility if applied correctly; it seldom is. Application of *extrinsic* motivation requires an in-depth understanding of human behavior.

AN ELDERLY GENTLEMAN

The following story illustrates the Multiplicative Principle

An elderly gentleman was bothered by the noise a group of young boys made when they played in an area near his house. The man tried several methods to get the boys to play elsewhere, but to no avail. Finally, he came up with a new and interesting strategy. He decided to pay the boys to play near his house! He offered them twenty-five cents apiece to return the next day. Naturally, the boys returned the next day to receive their pay, at which time the man offered them twenty cents to come the following day. When they returned again he offered them only fifteen cents to come the next day, and he added that for the next few days he would only give them a nickel for their efforts. The boys became very agitated, since they felt their efforts were worth more than a nickel, they told the man that they would not return! (Sidentop and Ramey, 1977)

The multiplicative principle suggests that the interaction between intrinsic and extrinsic rewards would either add to or subtract from intrinsic motivation.

*Caution: Leaders should realize it is possible extrinsic rewards can be counterproductive if the process is implemented improperly.

EXTRINSIC MOTIVATION

Unintended Consequences

Subjective Awards: Awards decided by opinions of an individual or group of individuals leaders.

Several years ago I observed an awards ceremony upon conclusion of a basketball program for kindergarten through fifth grade students. The motives of the men and women in charge of the program were honorable.

The application of the *subjective* awards produced unintended consequences. More harm than good was the product of the inappropriate application of awards: *extrinsic* motivation. More than half of the children returned home without a *subjective* award. The cronyism, nepotism and some politics was transparent. Regardless of age when an individual is treated unfairly, trust of the perpetrator is lost and trust most likely will never return.

The program's purpose was to create interest in basketball for future prospects. Mishandling the *extrinsic* motivation (*subjective* awards) produced numerous negatives such as a broken spirit or a feeling of inferiority. Future highly talented basketball individuals will migrate to another team sport or to individual sport where by its very nature the cronyism, nepotism and politics is a lesser factor.

Unintentional or intentional wrong always has negative consequences. Individuals in a leadership position who do not understand the *art of extrinsic motivation* should avoid it. This will take away from what could be accomplished. Far too often individuals in leadership positions have never engaged in formal training such as an academic class in mental skills, human behavior, or motivation.

Solution: In the preceding example this is what would have been appropriate: All the coaches line up and then have each participant's name called. At this point the young participant would go down the line of coaches and each coach give a positive comment with a high five. This proper award acknowledgment will stay with these participants for a very long time.

Once an individual recognizes the negative human behavioral traits which often filter into *subjective* awards; then and only then may they be properly addressed. In every instance of mishandling *extrinsic* motivation the results are counterproductive and the victims are the innocent. Leadership should understand the seriousness and importance of presenting awards in achieving motivation. From this example use transference to your profession in deciding how you may best use awards for your motivational environment.

The success or failure of Extrinsic Motivation depends upon two factors

- How you the leader implement (art) *extrinsic* rewards.
- How your team member perceives *extrinsic* rewards in their purpose of achievement

***Note:** The following comments concerning motivation in football programs at Penn State and Ohio State were written prior to both systems imploding. These systems collapsed because of failure to maintain the critical balance of *What Am I* with *Who Am I*. **We are all susceptible to this implosion.** *MSL* referring to *What Am I* with *Who Am I* as a critical balance is possibly an understatement. Without exception once the *What Am I* consumes the *Who Am I* the consequences are horrible, sad, and wide-spread. The longer an individual is in a leadership position, the greater the chances of this critical balance being violated, if not properly attended. In both of these football programs much success was achieved, but not its sustainability.

"IF IT AIN'T BROKE DON'T FIX IT"

Next time you get the opportunity to watch Penn State and Ohio State football, or other collegiate football teams, takes a careful look at their helmets. Keep in mind both teams just mentioned had highly successful team organizations.

Penn State helmets display no insignia. Ohio State helmets display insignia, indicating a performance goal reached. The goal the team member reached is a performance goal within their position. Penn State relies upon *intrinsic* motivation (no helmet insignia) while Ohio State uses *extrinsic* motivation (helmet insignia) to enhance *intrinsic* motivation.

Between the two highly successful organizations at Ohio State and Penn State, which one is correct? They both are. What fits the leader's personality should determine how *extrinsic* and *intrinsic* motivation is implemented within the organizational team structure. Would you use insignia or no insignia on team helmets? Putting yourself in these leadership positions, how would you use *intrinsic* motivation? If you are a potential leader, you may wish to focus on *intrinsic* motivation until you feel competent in the application (art) of *extrinsic* motivation.

Intrinsic and *extrinsic* motivations usage is not either or, it is the degree and ratio of both determined by your leadership. Great leaders in all professions use *intrinsic* and *extrinsic* motivation. How and when is totally dependent upon you the leader. Herein lies another wonderful leadership challenge in crafting motivation of best fit for your team members towards successful pursuit of your visionary goal.

In one's attempt to motivate, the greatest concern is not to destroy *Intrinsic Motivation*.

***Bottom line:** The most significant factor for team members concerning *intrinsic* and *extrinsic* motivation is: FAIRNESS.

Furthermore, before an individual can be motivated, that *individual must perceive* he or she has the necessary skills and ability to be competent in *achieving success* of their performance. Placement of a team member in the proper positional assignment is crucial in motivation. Placement of a team member in an improper positional assignment is detrimental to your *motivational environment*. Proper positional assignment is not static; it is fluid.

APPLIED: INTRINSIC MOTIVATION WITH CHARACTER

Outside the Cardinals baseball stadium in St. Louis, Missouri, stands a statue of one of the greatest hitters in baseball history, the stately figure of **Stan "The Man" Musial**.

Shortly after Stan Musial retired from his great career, an interviewer asked the following question, "Who was the best manager you ever played for?" Musial replied, "The one I was playing for at the time." Stan "The Man" Musial was highly, *intrinsically* motivated because he had self-confidence in his leaders, his team members, and himself.

Stan Musial was never thrown out of a major league game. However, it is my understanding he was tossed in the minors when his position, at that time, was a pitcher. Stan questioned a strike call which is automatic ejection.

STAN MUSIAL'S CHARACTER

As a small boy I attended a Cardinals baseball game in old Sportsman Park, St. Louis, Missouri. I was deeply impressed when the following game situation occurred: The visiting team was at bat. A long fly ball by the visiting team barely cleared the outfield fence, where Stan was playing. The umpire hesitated to make the call because he didn't know if a fan had touched the baseball as it cleared the outfield fence and into the stands. If a fan had touched the ball it would be ruled a double instead of a homerun.

The umpire was unable to make the call because his vision was partially blocked. The umpire called Musial over and asked if a fan had touched the ball. Stan informed the umpire a fan had not touched the ball. The ump ruled a homerun. On Stan Musial's word the umpire made a correct and important call favoring the opposing team.

What if You had been playing right field; the same circumstance happened and the umpire asked you the same question. What would you have told the umpire? Is it possible there is such a similar circumstance in your profession?

Stan Musial was a highly, *intrinsically* motivated team member, who participated with great character, on and off the playing field. Stan "The Man" Musial was extremely successful in maintaining the critical balance of *What Am I* with *Who Am I*. His character became eternal in my life and other lives he touched. However, of all the great records he accomplished I cannot name one. The *What Am I* is terminal in its importance; the *Who Am I* is eternal in its importance

Three Criteria for Motivational Environment

1. Successful Organization
2. Effective Communication
3. Performance Goals

1. SUCCESSFUL ORGANIZATION

Regardless of Profession each organization has two commonalities:

- **Leader:** A visionary with a strategy.
- **Team Members:** Those who pursue leadership's visionary goal.

Having the right talent in the right position at the right time is not a constant, but a dynamic variable. Herein lies a challenge for your leadership in creating your *motivational environment*.

***Important**: Position leaders and team members are people. Successful leadership always comes back to people, and how you the leader treat them. Without exception, your professional relationship with your team members will determine your degree of success or failure. Providing a *motivational environment* is an immeasurable benefit in pursuit of all goals within your organizational team structure.

THERE ARE THREE CATEGORIES OF ORGANIZATIONAL TEAM STRUCTURE: SIMPLE, MEDIUM, COMPLEX

Simple Organization: Leader with one or several team members.

Medium Organization: Leader with positional leaders and many team members.

(Position leaders are team members)

Complex Organization: Any large industry such as airline industry, retail industry, automotive industry, oil industry and other major industries have a complex organizational team structure. Many complex organizations are multi-national. These industries have a CEO and many levels of position leaders with thousands of team members.

In your profession how would you classify your organizational team structure?

Simple

Medium

Complex

The Organizational Structure in Military and Sport has been developed to a near perfect science. In the military rank and order are engraved in stone. In the profession of team sport the organizational structure is predetermined by position.

However, the secret to successful organization is totally dependent upon your leadership ability, to place the right team member in the right position at the right time: The glove of best fit. Having done so will greatly accelerate and increase the success of your motivational environment.

In the profession of sport there is full disclosure and transparency in every team member position and leadership position of the organization. Successful leaders place team members and position leaders in the correct positions at the correct time, another exciting challenge for your leadership ability. If these adjustments and challenges were not necessary, your leadership position could be replaced by a computer.

It is necessary for your team members to be assigned the correct positions at the correct time. In baseball and softball the leadoff hitter is determined by who will most consistently get on base. What if your leadoff hitter goes into a hitting slump? Is it the right time to make an adjustment in the batting order? Do you have the ability to make this adjustment without affecting the self-confidence of your leadoff hitter, while maintaining a positive team culture? Great leaders approach these and other challenges with courage and passion.

What about your team? Do you have the right team members in the right positions? Placing the correct team members in the correct positions is a necessary, delicate process for developing and *sustaining* your successful *motivational environment*.

ART OF MOTIVATION: APPLICATION

2. EFFECTIVE COMMUNICATION

Effective Communication Defined: When you the sender (leader) *create a duplicate picture* in the mind of the receiver (team member).

In many professions *effective communication* may be comparable to a trout stream that has a smooth stream, ripples, pools, and rapids. However in military, first responders, and sport, *effective communication* is likened to a roaring river. *Effective communication* is vital in all professions to enable a *motivational environment* for team members to successful pursue your visionary goal.

It is incumbent upon leadership to create a duplicate picture of your visionary goal in the mind of each team member. This duplicate picture may only be verified by proper feedback from leadership with each individual team member. (Chapter 4 Goals with *Feedback*)

Effective communication is a two way street. Research surveys indicate as much as 97% of all issues, conflicts, and problems within a team structure are the direct result of ineffective communication in both profession and life. The remaining 3% may be attributed to personality conflict, burnout, and other such factors.

QUESTION: WHEN A TREE FALLS IN THE FOREST

When a tree falls in the forest and no one is around, does the fallen tree make a sound? No it does not. It only becomes a sound when the sound waves are recorded by a receiver. No receiver, no sound, only sound waves. The same holds true in your communication with team members.

Never assume anything in communication. It is your leadership responsibility to create and verify the duplicate picture in the mind of your team member. Determining if the duplicate picture has been completed is one of the functions of feedback. *Effective communication* is vital in developing and sustaining your *motivational environment*.

EXAMPLE: EFFECTIVE COMMUNICATION

WEST POINT VOLLEYBALL

I Think They Do

A good friend of mine, who has a genius I.Q. and an eidetic memory, was the head volleyball coach at West Point in which there were 15 team members. In our conversation of mental skills, I asked my friend if each member of the squad knew what he expected of them in their position. He remarked, "Well, I think they do."

He offered himself this challenge: Write down what he expects from each team member in their respective position. Then have each team member write down what they think he expects of them within their team position. He did so. Three team members of the 15 member squad were on the same page the coach had written down. Feedback from each team member exposed the need for improved communication.

Once discovered, great leaders successfully address this communication need. The West Point volleyball coach successfully addressed improved communication. The discovery of these needs was the product of *effective communication* with *Feedback*.

***Caution:** Effective Communication is a Two Way Street. Surveys strongly indicate that leaders are far better in sending communication than leaders are in receiving communication. In your leadership role, open mindedness and composure in receiving communication from team members (feedback) is a vital asset in developing your *motivational environment*.

EFFECTIVE COMMUNICATION EXPERIMENTS (2)

Sing a Song

In a Western state, the following exercise was conducted during a seminar with the twelve position leaders singing a different song simultaneously. Had the windows been left open in the cabin there is no doubt the bears, antelope, deer, possums, raccoons, squirrels, mice, bobcats, pumas and all other woodland critters would have fled Paradise Valley, with some never to return.

If you have 12 team members (this can be any number) in the same room place the members in a circle. With larger numbers incorporate as many circles as necessary. Give each team member different written song lyrics such as Mary Had A Little Lamb, Old McDonald Had a Farm, etc. Have each of your team members sing at the same time. Not a pretty sound to the human or animal auditory process.

Next, each individual would be given the same lyrics and would sing simultaneously ... much better effective communication. Every team member being on the same page enables effective communication and greatly improved motivational environment.

It is incumbent upon your leadership to verify each team member being on the same page. Your motivational environment requires *effective communication*, creating a duplicate picture. Ineffective communication is never a pretty thing and always results in unnecessary conflicts and issues.

Grocery List

Verbal with No Feedback: Any number of team member participants is acceptable. However, for this experiment we'll use 12 to 15 team members sitting side by side in a circle or semi circle. The leader starts the experiment by showing a team member a grocery shopping list of three items (leader keeps the written list). Example: 2 loaves of bread, 1 pint of skim milk, 6 bananas.

Now your team member whispers the three item grocery list to the next team member. This process is repeated until it has been given to each team member. When the circle is completed, it will finish with the leader who originally started the grocery list. The final grocery list will not be the same as the original. Never assume a duplicate picture in communication.

Verbal with Feedback: Repeat this experiment with one exception--feedback. Each time, the sender of the grocery list requires the receiver to give verbal feedback to verify a duplicate picture. When the grocery list completes the circle it will most likely be a duplicate of the original list. *Effective communication* requires feedback for a duplicate picture.

Three Types of Communication: Verbal, Paralanguage, Kinesis

Studies indicate the impacts of communication between two individuals are as follows:

- **7% Verbal:** Spoken words. The world has thousands of different languages. Regardless of language, when spoken it is classified as verbal.

- **38% Paralanguage:** "Paralanguage is also spoken. However, it is not what you say, but how you say it." (Martens, 1997, p. 59) Some of the components of Paralanguage: Tempo-how fast or slow, Volume-how soft or loud, and Rhythm-how much emphasis placed on different words and the Cadence of speech.

- **55% Kinesis:** Body language–How you communicate through physical appearance or by your demeanor. (Non verbal) A few Components of Kinesics: Gestures, Different facial expressions, Eye movements and Physical Demeanor

KINESICS KNOWN WORLDWIDE

Smile

Planet earth has a population of seven billion, approximately 200 countries, and 6500 languages; each individual of this demographic understands the meaning of a common kinesics: the smile.

Gesture

Another example of kinesics (gesture) that gained worldwide notoriety is the product of the captured crew of the *USS PUEBLO* January 23, 1965. A technical research ship (Navy Intelligence) the *USS PUEBLO* was captured off the east coast of North Korea and its crew was imprisoned.

The captured Americans were tortured. However, their captors denied any wrong doing, and insisted the crew was being treated humanely. Our entire nation was greatly concerned if the crew would survive their imprisonment.

North Korea, in attempting to falsely verify the crew being treated fairly, decided to take a picture of the *PUEBLO* crew, clean shaven and in their military uniforms. This picture was taken and released through all the major news media around the world. The captured crew knew this may be the only opportunity of letting their loved ones and the free nations of the world know they were indeed being tortured. At the moment the camera took their picture, each U.S. crew members used kinesics. Oh yes, they did; each used the hand gesture of "flying the bird." At this point their captors were unaware the meaning of this kinesic gesture.

The American crew made full disclosure of this ill fated attempt to prove to the world the prisoners were being treated fairly. These brave Navy warriors knew when their captors learned the meaning of this gesture, they would be severely punished; and so they were. Eventually the crew returned to America. Today 2013, the *USS PUEBLO* remains a hostage on the Taedong River in North Korea. (Ring, 2008)

Because of this international incident, "flying the bird" is a kinesics understood by the world. Move over smile. This kinesic demeanor by the crew of the *USS PUEBLO* was an enabler for *effective communication*.

Your leadership understanding of the three methods of communication is a positive building block for creating your *motivational environment*. Hopefully the kinesics gesture you will experience most often is the smile.

3. PERFORMANCE GOALS

Individual team members' performance goals enable discovery of strengths and weakness. Understanding strengths and weakness is a major component in developing a motivational environment.

Strengths: Ability to successfully accomplish performance expectations

Weakness: Inability to successfully accomplish performance expectations

Providing performance goal setting will be the heart and soul of your *motivational environment*. Regardless of profession, performance goals reveal individual team member's strengths and weaknesses (needs).

Mental, Physical, and Technical skills are the major categories in which strengths and weaknesses occur. Team members will have one, two, or all three of the following needs within their positional assignment:

* **Technical:** Specific trained skills needed for successful positional expectations
* **Mental:** Goals, Refocus, Motivation, Stress (relaxation), Self-confidence
* **Physical:** Endurance, Strength, Stamina, Power, Dexterity

By its very nature, performance goal setting reveals and addresses your team members' needs within their assigned positional boundaries.

Chapter 4 discusses in detail setting proper team member, *fail-proof* performance goals.

Developing performance goals within your motivational environment requires three leadership skills

- **Teaching** (Educating)
- **Counseling** (Listening)
- **Coaching** (Implementing)

Simple Organizational Structure: If your organizational structure is a simple structure you most likely have by default inherited the role of teacher, counselor, and coach.

Medium Organizational Structure: If your organizational structure is medium, you may find it necessary to assign position leader(s) with expertise in teaching, counseling and coaching.

Complex Organizational Structure: If your organizational structure is complex you most likely have positional teachers, positional counselors, and positional coaches. One other effective choice is to acquire professional experts in this field to assist in the implementation of a team member performance goal setting program. However, it is preferable you first look within your organization for positional leaders who qualify in these three areas.

*As leader, it is greatly beneficial to fine tune your teaching, counseling and coaching skills to evaluate the effectiveness within your team structure. If you have position leaders in your organization it is equally important for them to refine these skills. Great Leaders are good teachers, counselors and coaches. The opposite is also true. Great Teachers are good counselors and coaches. Just as the mental skills of goals, imagery, relaxation, and refocus are interrelated so too are the leadership skills of teaching, counseling and coaching interrelated.

Developing a *motivational environment* for team members is a three stage process: *Teaching* (Educate), *Counseling* (Listen), and *Coaching* (Implement). This process may be accomplished in team meetings and one on one.

Teaching Defined:

Effectively COMMUNICATING vital information to team members with enthusiasm and passion

Counseling Defined:

Issue and conflict resolution requires *GUIDING* team member to discover the answer; when counseling you are a Peaceful Warrior, *LISTENING* with an open mind. *Counseling* does not give direct advice. *Counseling* provides strategies for best solution. (Hayward, 1998)

Coaching Defined:

Ability for *UNDERSTANDING the heart and mind* of team members thus enables identifying team member's *strengths and weaknesses.*

TEACHING

VINCE LOMBARDI: THE GREAT TEACHER

The actor portraying Vince Lombardi in the Lombardi movie studied this renowned leader in great depth. This actor made the following comments while being interviewed on national television prior to the film release, "Lombardi considered his role as teacher with as much importance as he did his role of coach. Vince's goal surpassed just winning; his goal was to develop the highest level of performance in each of his team members." Vince Lombardi knew the importance of *Teaching*.

First and foremost, great leadership must provide good teaching. (Williams, 1986) The absence of good teaching is a distinct disadvantage. Providing a teaching environment is necessary to develop your *motivational environment* to its full potential, regardless of organizational structural and size.

Your teaching environment should include:

* Caring, cultivating, and developing the talents of each team member
* Teaching each team member and helping him or her to clearly understand their positional boundary
* Teaching each team member their positional expectations within their positional boundary and how their assignment may be accomplished

*Continually reinforce leadership's visionary goal. Your visionary goal should transcend your *motivational environment*.

Qualities of Teaching Excellence

Passionate..Entertaining.. Humorous..Listening..Mentoring..Counseling.. Coaching. Each quality is beneficial in developing intrinsic motivation. What about your passion? Does your energy level sky rocket when you are engaged in your leadership performance?

Teaching Methods (2)

1. **TELL UM METHOD**

    Tell Um what you are going to tell um

    Tell Um

    Tell Um what you told them

May sound simplistic but it is highly effective. The Tell Um Method is effective in developing speeches, presentations, writings, team meetings, etc.

2. KISS METHOD: Keep It Simple Stupid

While attending the United States Sports Academy, I heard many of my Southern colleagues referring to the KISS Method. I finally got up enough courage to ask "What does the KISS acronym stand for?" Answer, *Keep It Simple Stupid*. They all seemed shocked I was unaware of its meaning. At that particular time the KISS method was pretty much a "Southern Thing." Incorporating the *KISS Method* with the *Tell Um Method* is a formula for a pleasurable and a meaningful experience.

Reminder: Teaching is *effectively communicating vital information* to team members with enthusiasm and passion. Effective teaching is an important building block for *trust*.

A very large majority of us humans are visual learners. Appropriate visual aids are extremely beneficial in *effectively communicating* information. In teaching excellence, verbal interaction between teacher and student allows information a longer duration in the human brain's memory process. Question and answer interaction is a major benefit.

COUNSELING

The Peaceful Warrior

Foundation of Counseling: Listening with an open mind.

One of the numerous benefits of counseling is increasing self-awareness. Integrating counseling skills within your leadership structure is a necessity for creating a sustainable *motivational environment* in pursuit of your visionary goal. Counseling requires careful *listening* for one's *needs*, then providing a menu of suggestions to enable a team member or a team towards discovery of solution. All great counselors are great listeners. When you develop listening skills you create an environment for greater accomplishment. In counseling you assume the role of a Peaceful Warrior. You are not telling. You are *listening* and offering options for successfully fulfillment of a need or needs. Research indicates many leaders in all professions are far better in telling, than they are in listening. As will be discussed later, counseling is vital in assisting team members to develop proper performance goals.

JOHNNY CARSON: THE GREAT LISTENER

Why was this great host/comedian so successful? How was he able to set a standard in his profession that has eluded so many others? He developed the skill of listening to a higher level. What was Johnny's purpose in listening? His awesome listening skill enabled him to achieve getting his guest (team members) to display their best talent at that particular time. Many others have failed in Carson's profession because they falsely believed the show's performance was about them and not their guest (team members).

Is there a lesson here for leadership in any profession? Is it possible some leaders falsely believe it is about them and not their team members? Listening is not just a counseling skill; it is a necessary skill for elevating leadership to its highest level. Johnny Carson developed the art of listening to its maximum level. Is it incumbent upon each of us in our leadership role, to pursue the same quality of listening?

The science of listening is developed not just within the mind, but also within your heart. Passionate caring about your team members increases the progress towards the successful development of your *motivational environment*.

EXAMPLE: BAD COUNSELING

"WHAT? Are you kidding me!?"

Early in my educational career, as athletic director, my office shared a very thin wall with one of the counselor's offices. One morning a young male student entered the counselor's office for his scheduled appointment. The following conversation occurred between this highly educated counselor and a student we'll call Paul. "So Paul, I see over the last several weeks your grades have been slipping in each of your classes. Is there a reason for this?"

After a semi-long pause, Paul reluctantly and solemnly answered, "Well, my dad has been drinking lately, and he has been beating up on my mom."

The counselor's response, "What does that have to do with it?"

Paul had an immediate, elevated verbal response (paralanguage), "*WHAT?! Are you kidding me?*"

"Oh, well I guess so," replied the counselor. No matter what the profession there will always be someone who slips through the cracks. A large percentage of counselors perform with excellence.

Developing your listening skills will become a trait of your character. Team members evaluate your leadership on the content of your character. Likewise, you will evaluate team members on the content of their character. Your relationship with your team members will determine the failure or success of your visionary goal. Listen carefully with an open mind.

COACHING

Coaching requires understanding the *Who Am I* of those you serve. Coaching is a vital skill of every great leader regardless of profession. Great leaders are great coaches. Your leadership coaching ability is an asset in developing and sustaining your *motivational environment*. Leaders should be aware your team members are people who have feelings and needs. Your sustainable success will always depend upon how you address your team member needs. Coaching enables team members to identify their personal strengths and weaknesses (needs). Assisting team members to properly develop goals is a duty and responsibility of leadership coaching. Coaching is the ability to get inside the heart and mind of team members.

Four Valuable Coaching Virtues

- Confidentially
- Trust
- Composure
- Fairness

CONFIDENTIALLY

Successful coaching has certain criteria that must be met. *Confidentially* is certainly one of the virtues. Publically exposing or criticizing a team or a team member's goal will never provide positives. *Confidentially is critical.* Violating confidentially causes permanent distrust between two individuals or between leadership and team. Coaching should stress to team members each one carries responsibility of *confidentially*.

Leadership coaching is an integral part in developing proper goals for team members. Goal setting techniques enable a relationship between leader and team member to clearly identify each team member's responsibility towards other team members and leadership. Your professional coaching relationship is necessary in developing your *motivational environment*.

Successful coaching enables everyone to be on the same page. Stated earlier, a team member who doesn't understand the whole vision won't understand their piece of your vision. Successful leadership coaching involves clear communication interwoven with *confidentially*.

TRUST

Trust and confidentially are as one. If one is violated, both are violated. When leadership publically exposes and or criticizes a team member's goal, leadership has lost the trust of that team member. *Trust* is a trait among all great leaders. Developing a trusting relationship between leadership and team members is a virtue with numerous benefits.

EXAMPLE: CONFIDENTIALLY AND TRUST VIOLATED

Certifiable Psychotic Division I Football Fan

Recently I was asked by an avid and, yes, possibly psychotic Division I football fan, "What has gone wrong with our football team this year?" The Division I University football team she was referring to had only lost one game the previous two seasons. This year the highly successful team had lost two games by mid-season with practically the same team members. Before I could respond to her question she continued, "No one knows the answer to that."

Not passing up this opportunity I responded, "Well I know the answer." Of course her reply was, WHAT!?(Paralanguage)

"Do you remember the fifth game this year when your team was down 17 points going into the fourth quarter? They mounted a near impossible comeback on the opponent's field and won by two points on a field goal with time running out. Well, after the game your coach (leader) made a common and costly mistake in his post game interview."

When the sports commentator asked your coach about his team's great comeback he made this hubristic comment, "I am not pleased with our kids performance at all; we played poorly to get that far behind." Trust between leader and team members was greatly compromised. During sports seasons listen carefully, you will most likely discover the leadership of others with this hubristic violation of trust on a frequent basis.

COMPOSURE

Instead of publically complementing his team's great comeback, he allowed "Ole Huber" to violate the trust and confidentially between leadership and his team members. Keep your *composure*. The remainder of the season the team never played up to their' expectations. How would anyone of us feel if we were in the shoes of those team members?

One might ask, "Didn't the coach have the right to express his feelings?" Answer, no and yes. **No**, leadership should never be critical of a team or team member in a public forum. **Yes**, behind closed doors. Never air dirty laundry in public. This will create a downward spiral making it difficult to apply the brakes.

Developing trust and confidentially in your leadership will always produce numerous benefits. This seasoned and famous coach was blindsided by a hubristic attitude. Without exception when one loses composure, "Huber" will be present tempting your acceptance. Keeping your *composure* when emotions are high is not an inherent human behavioral trait. One must acknowledge this human weakness in attempting to develop your skill of *composure*. (Could this possible be a performance goal?) *Composure* is a very close relative of focus.

CLASSIS ICON OF COMPOSURE

AARON ROGERS

Green Bay Packers Quarterback

The following situation occurred after the sixth game of the 2012 NFL season. The Green Bay Packers won/loss record was two and three before the game against the undefeated Texans. Both local and national news media had been publically criticizing the great veteran quarterback, Aaron Rogers. The Packers thumped the undefeated Texans with Rogers throwing six touchdown passes.

In the post game television interview the final question from the young female reporter to Rogers, "What would you like to say to your recent media critics?" Aaron Rogers' response is a classic in *composure*; applying his forefinger in front of his lips, **"Shhhhhhh."**

Then he immediately exited the interview to be with his team mates ... an icon for keeping one's *composure* in public. If you had been in Aaron Rogers shoes how would you have responded? It's very beneficial to keep your criticism in private and certainly your *composure* in public.

How do you reason with an Idiot

Great leaders often develop a mental dialog for composure. The following true story is an example:

Our basketball coaches were in a post game staff meeting after having lost a close game to our rival school. The staff meeting was abruptly interrupted when an irate uptown coach burst into the coaches' room. The uninvited guest walked up in front of the large desk where the head coach was sitting. Our head coach was an enormous physical specimen of a man. The tall, scrawny uptown coach verbally insulted each of the staff. The only comments our head coach made in the one sided exchange was, "You're probably right; you're probably right."

Eventually the uninvited individual left. An assistant coach immediately asked the head coach, "How did you keep your *composure*? How did you keep from jumping across the desk and clocking him?" His reply, *"How do you reason with an idiot?"* Developing an internal dialog enables you to keep your *composure* under difficult circumstances. (Chapter 7 Refocus)

Composure is not a human behavior which is inherent, but a behavior acquired through deliberate, self-disciplined focus in emotional situations. Understanding this weakness and taking it to heart is acknowledgement of your understanding the value of art and science of human behavior. Developing a preplanned internal dialog is an invaluable technique in maintaining *composure*. Writing down your preplanned composure response will imprint your response in the brain's memory. What would be your preplanned response for *composure*?

Fairness

Total *fairness* is impossible. It is your sincere effort towards *fairness* which is valued by individual team members and consequently throughout your team culture. In your professional leadership you have discovered or will discover 99.9% of all you do is a process, not an event; so it is with *fairness*. Great leaders see *fairness* as an exciting personal challenge.

Confidentially, trust, composure and *fairness* are necessary traits in elevating your leadership to its highest sustainable level. These traits will provide numerous positive results during your leadership's tenure and will develop lifelong values for many of your team members in both their professional and personal life.

*****Caution:** A leader who compliments a half hearted performance, will damage his or her character in the eyes of team members. Complimenting half hearted performance is unfair and detrimental to sustaining your successful *motivational environment*.

Your leadership will always have lasting values for your team members when you develop a *motivational environment* including *teaching, counseling, and coaching*. It is possible as leader you may never know the positive impact your values have made on those you have served and on those who have observed. Positive characteristic traits are eternal in your legacy.

GREAT LEADERSHIP'S LASTING VALUES: LEGACY

Following is a letter to the editor of a newspaper 58 years after the fact.

Thank You Coach John Evers and Coach John McDougal

The years I reflect upon are 1955-1959 Carmi High School. I played basketball all four years, track and field my junior/senior years and baseball my senior year. I played sports because I wanted to play for these coaches. What was important to them was building character and mental toughness.

My senior year in basketball I set the school record for points scored and was unaware of that record until several years later when it was broken. I set the 100 yard dash record and was unaware of that until I saw the trophy while attending a Carmi basketball game several years after I had graduated. You see, records were not important; how hard you participated and how well you conducted yourself on and off the playing field was important.

While in college, one of our favorite past times was getting together to play basketball. After we finished we would sit around and tell tall tales of our high school days. Often times players would lament that if they just would have had a good coach in high school they would have accomplished great things. After they all finished I would make the following comment, "You know those coaches you wished you had; well I had them."

One day in practice our drill was shooting layups in which there were three lines of players and Coach Evers was rebounding. It was my turn next in the middle line. While coach had his back turned rebounding, my teammate dropped my trunks below my knees. Oh yes, the GAA girls were on deck practicing. While everyone was laughing, I quickly pulled up my trunks.

Coach Evers turned around and said," Sam, you are the only one not laughing so I suppose you are the one causing the problem. Go give me two miles on the track." Coach knew I was a sprinter and hated distance; plus it was cold and dark. Having been so unjustly persecuted, I ran the fastest two miles I could possible run.

Coming back into the gym sweating and proud, I saw Coach Evers look at me and heard him say, "You haven't had enough time to run two miles. Go give me another mile." Back at the dark cold track I proceeded to run the slowest mile ever recorded in the history of planet earth.

Why did Coach Evers have me do that? He wanted to know if I would follow orders and he wanted to know how mentally tough I was. That is what was important.

Being the typical 17 year old, I started heading down the wrong path of life. Coach McDougal and Coach Evers called me in to what many would refer to as a "Come-to-Jesus meeting." Boy did I get it, and boy did I have it coming. Every day of my life I am so thankful that my coaches cared enough about me to take time out of their busy schedule to put me back on the right track. Character building was extremely important to each of them.

The late Coach John Wooden said, "Who you are as a person is far more important than what you are as a basketball player." My coaches spoke these words from their hearts each day to me and my basketball teammates.

Yesteryear, Coach Evers, United States Marine, served in WW II and the Korean War while Coach McDougal served in WW II. Today, in their senior years, both coaches have very difficult health challenges, which they approach with character and mental toughness.

Coach Evers and Coach McDougal were my teachers, my counselors, my coaches, and my heroes. Thank you Coach John Evers. Thank you Coach John McDougal.

Addendum: Several months after this letter to the editor Coach John Evers succumbed to Parkinson's disease. This is a reminder we only get one chance at life with good character and each of us live on borrowed time. Regardless of profession and life circumstance, it is important we enjoy, do our best today and each tomorrow. Great leaders produce lifelong benefits for the lives they touch.

YOUR LEGACY

How will you want your team members to reflect on your leadership values? When the day comes and you cross the finish line, what will you have given to those you served? Will some team members feel you made a difference in their life? How will you want your team members to remember you after your career exit? Evaluating your perception to this important question with a systematic review is an enabler in keeping your critical balance of *What Am I* with *Who Am I*. **What will be my**

legacy is a thought in the process of developing your visionary goal for leadership in profession and personal life.*__Remember:__ Achieving successful, sustainable leadership is all about team members; it is about your professional and personal relationship with team members.

SUGGESTION: OH NO

Enroll in a university or college class in counseling, coaching, or teaching. This experience will be beneficial and exciting. Being on a college campus is a unique experience you will not regret. Age is not an excuse. You may even ask yourself: Why haven't I done this before. The positives of this experience will be breathtaking.

HOW TO ENROLL IN COLLEGE

Suggested process if you wish to consider enrolling in a college class: Contact the head of the college department of your interest. Ask the secretary to put you through to the department chair. He or she will be more than happy to discuss your interests and will provide you with insights to fulfill your needs. In many instances you may audit a class if you so desire instead of taking the class for a grade. Suggested Departments: Counseling, Education, Human Performance, or Psychology

* Ask about the class professor. Without exception what makes a class a positive and enjoyable experience is the professor (leader).

How Our Brain Works: Do not be afraid of the suggestion to go back to college. In our quest to learn, scientists have discovered the three-pound, walnut-shaped gray matter, the human brain goes through the following stages for maximum learning:

- In fetal development
- After birth
- Between the ages of 4 and 12

- Anytime during the remaining years of your life when properly exercised

"The brain learns and remembers throughout life. It employs the same processes it uses to shape itself in the first place; constantly changing its network of trillion connections, opening and shutting gates between and within the different structures of nerves. All this is a result of stimuli from your brain's environment. The brain is very capable of rewiring itself. However, what the brain can do depends on whether or not it is used. The brain is eager to learn new skills." (Doidge, 2007, pp. 6-8) Seriously consider a college class to sharpen your leadership skills. In leadership, *Hope is Not a Method*.

False Justification: Some individuals may have false justification for not improving teaching, counseling, and coaching skills by convincing oneself I don't have enough time, it is beneath me, or it's simply a bad idea. Translated, I'm scared. Fear is a normal human behavioral reaction. One of the three characteristics of great leaders and performers: **they do not fear failure**. They actually accept new challenge to evaluate their ability for future success.

Knowledge is the Antidote to Fear
RALPH WALDO EMERSON

Example: False Justification put in its Proper Place

FLAMBOYANT ATTORNEY

A very flamboyant attorney, while questioning the expert forensics witness in a Southeastern state in the U.S., asked the following question, "Even though the victim's head was almost severed, isn't it possible he still could have been alive?"

The expert witness responded as follows. "Yes, I suppose it is and I suppose somewhere in this state there is an attorney practicing law without a brain." Everyone in the court room appreciated this analogy; that is except the judge. Remember how the human brain works. It can justify anything, a headless body still alive and even murder.

Great leaders are always seeking ways to pursue and achieve excellence. Seriously consider and evaluate your leadership providing effectiveness in teaching, counseling and coaching. It is very important for creating full potential of *Your Motivational Environment*.

WEST POINT INTERVIEW

While at West Point I interviewed the Colonel who developed the world renowned Performance Enhancement Training Center at this incredible Military Academy. I presented the Colonel with the following scenario: "Right outside your office I am going to bring three buses of young men and women who desire to become successful leaders. They want to know from you one thing you believe would help them in pursuit of successful leadership. What would you tell them?"

After some thought he spoke. "I'd tell them to find out what it is you want to achieve then don't let anything get in your way of this accomplishment (your goal), while staying true to yourself (your character)." The critical balance of *What Am* I with *Who Am I*.

In this same interview, I also asked the Colonel, who served in several wars including Viet Nam, "What is the hardest thing you have ever had to do in your leadership role?" His facial expression changed as did his entire demeanor (kinesis). Solemnly he spoke deep from his heart, "Notifying parents their son or daughter had paid the ultimate price." He then leaned forward looked me straight in the eye with a stone face and said, "Here at West Point we have a saying, *Pick and Choose What You Fall on Your Sword For.*" Translated: There is a price you must pay to become a great leader.

Elevating your present leadership to a higher level requires a disciplined, confident, and intentional effort. Previously, *MSL* related the U.S. Military now trains with goal setting performance. One of our most highly decorated and successful Military Generals testified before a congressional committee during the second Iraq war. A couple of congressional committee members (both genders), trying to score political points, inferred this war hero was a liar. This General never batted an eye. How did he keep his composure?

- He had no control over their comments.
- He was focused entirely on his performance goal in his pursuit to accomplish his leadership visionary goal in this military theater. (Chapter 7 Refocus) Not only was this military hero's leadership visionary goal successful, it also was credited for a successful ending to our mission in Iraq.

Using mental skills in your leadership role, you will always have your critics. These individuals can be compared to the hubristic behavior of the mentioned congressional committee members. Your successful leadership requires you keep your composure in staying true to your visionary goal. Your composure is necessary for your team members to stay true to their performance goals. You are their leader; you are their role model. Your words and actions are affirmations for team members doing the right thing.

When Your Leadership Develops

Successful Organization

Effective Communication

Performance Goals

You have created the perfect storm for a team culture with an effective *Motivational Environment* in pursuit of your visionary goal.

Four Issues to be addressed in Your Organizational Structure:

- Public Criticism
- Nepotism

- Revenge

- Self-Discipline

Public Criticism: Public Denigration

A leader's public criticism of any personal performance is a monumental negative factor. Leader's public criticism destroys team member's trust in your leadership. **Never, Never, Never** publically criticize a team member or a team in a public forum. Keep your composure. Your leadership ability to maintain composure during an emotional environment greatly increases team member's motivational development. Your words are validated by your actions and your actions validate your character. Composure is a necessary character trait for elevating your leadership to its highest sustainable level.

Nepotism: Favoritism; Bias

In your leadership organization there is no place for nepotism. Just as there may be a good ole boys club, at times there may also be a good ole gals club. Nepotism knows no gender and has no positive qualities in developing a successful motivational environment. Nepotism is elusive and malignant with a final outcome of eventual total destruction of your leadership.

Revenge: Reprisal; Pay Back

Never seek revenge. (Oh, What a Temptation.) There is no escaping a team member doing or saying something which was improper or hurtful, whether intentional or unintentional. It is normal human behavior for your ego to silently shout, "I'll make him or her pay for that." Revenge is a poison. Leadership revenge destroys *motivational environment* and speaks volumes about character. Revenge is a human behavioral trait great leadership resists with success.

Understanding the science of human behavior is a major plus in understanding the positives of eliminating revenge in your leadership role. When you seek revenge you are allowing the other person to live rent free in your head. Are there any positives of revenge in leadership or personal life? **Proverb:** Before you seek revenge dig two graves.

Self-Discipline: Personal Control

The importance of self-discipline is found in the following words of Thomas M. Sterner (2008). "Of all the riches available to us in life, self-discipline is surely one of, if not the most valuable. *All the things worth achieving can be accomplished with the power of self-discipline.* With it we are masters of the energy we expend in life. Without it we are victims of our own unfocused and constantly changing efforts, desires, and directions."

Carpe diem-seize the day: Self-discipline, focus, patience, and self-awareness are interwoven threads in the fabric of true inner peace and contentment in life: Together, living in the present moment and being process-oriented is the path that leads us to these all important virtues. (Sterner, 2005)

Self-discipline is a constant stream of refocusing circumstances. Choosing not to refocus will result in loss of self-discipline and eventual failure of potential great accomplishment. Self-discipline will empower you on your journey to a higher level of sustainable leadership in profession and life.

*Failure to address these issues will greatly compromise your *Motivational Environment.*

AN **INDIVIDUAL** WITH A **VISIONARY GOAL** THAT HAS THE ABILITY TO **MOTIVATE TEAM MEMBERS** TOWARDS ACCOMPLISHING **YOUR VISIONARY GOAL**

CHAPTER 3 MOTIVATE TEAM MEMBERS
CREATING A MOTIVATIONAL ENVIRONMENT

SUMMARY

Perception of Success is a Precursor to Motivation

Chapter 3 presented *creating a motivational environment* for those you serve. All your successes and failures will originate from how you treat your team members. Your knowledge of human behavior (Science) and its application (Art) is necessary in creating a successful and sustainable *motivational environment* for team members.

CHAPTER 3 MOTIVATE TEAM MEMBERS: MAJOR TOPIC POINTS

SCIENCE OF MOTIVATION: KNOWLEDGE

Two Types of Motivation

> Intrinsic–from within
> Extrinsic–outside source

Two Theories

> Additive Principle (additive not interactive)
> Multiplicative Principle (interactive)

Three Criteria for Motivational Environment

> Successful Organization
>
> Effective Communication
>
> Performance Goals

ART OF MOTIVATION: APPLICATION

Effective Communication

> Verbal–7%
>
> Paralanguage—38% (how you say it)
>
> Kinesis–55% (non-verbal: smile, gesture)

Performance Goal Setting Environment

Enabler for Your Discovery of Team Member Strengths (Abilities)

Enabler for Your Discovery of Team Member Weakness (Needs)

Leadership Skills to Consider

Teaching–presenting vital information with passion

Counseling–A Peaceful Warrior

Coaching—ability to get inside the heart and mind of team member: the Who Am I

Issues to be Addressed in Your Motivational Environment

Public Criticism-never

Nepotism-serves no positive purpose

Revenge- poisons team

Self-Discipline-a close relative of composure: Shhhhhhh

CHAPTER 3 GLOSSARY: MOTIVATE TEAM MEMBERS

Additive Principle: The notion that intrinsic and extrinsic motivation combine to create need achievement.

Coaching: Ability to UNDERSTAND the heart and mind of team members thus enables identifying team member's strengths and weaknesses.

Carpe diem: Seize the day

Competence: Competence is a team member's conviction for a successful outcome

Counseling: Issue and conflict resolution solved by GUIDING the team member to discover the solution

Effective Communication: When leadership and every team member are on the same page

Eidetic: Mental images, unusually vivid and almost photographically exact

Environment: A carefully crafted atmosphere that provides boundaries from which a team member may pick and choose from a selected menu of acceptable stimuli (mental techniques) designed for a specific expectation which influences behavior and outcome.

Extrinsic: Motivation that comes from an outside source. External rewards to motivate behavior in performance such as, money, ribbons, medals, trophies, and praise

Intrinsic: Motivation that comes from within an individual (team member) Motivation to participate in an performance for its own sake and for no other reason

Motivation: The degree of mental intensity directed towards the accomplishment of a goal

Multiplicative Principle: The belief that intrinsic and extrinsic motivations are interactive and not additive

Nepotism: Favoritism; Bias

Public Criticism: Public Denigration

Revenge: Reprisal; Pay Back

Self-Discipline: Personal Control

Subjective Award: Award decided by opinions of an individual or group of individuals leaders.

Team: *E Pluribus Unum* "Out of many, one"

*Creating a Motivational Environment is both a
Science and an Art*

AN **INDIVIDUAL** WITH A **VISIONARY GOAL** THAT HAS THE ABILITY TO **MOTIVATE TEAM MEMBERS** TOWARDS ACCOMPLISHING **YOUR VISIONARY GOAL**

PART I:
SUMMARY

◆ ◆ ◆

SCIENCE AND PHILOSOPHY OF YOUR LEADERSHIP

Proverb: *The strength of the Wolf (you the leader) is in the Pack (your team members); the strength of the Pack (your team members) is in the Wolf (you).*
RUDYARD KIPLING

Chapter 1 An Individual: Focused on self-evaluation for developing your personal *Philosophy of Life* and your *Philosophy of Profession*. Furthermore, your Critical Balance of *What Am I (your profession) with*

Who Am I (your character) was presented stressing importance of this balance and the inevitable destruction when this balance is ignored.

Chapter 2 Visionary Goal: Presented a strategy to develop a clearly articulated *Visionary Goal* transcending your organizational structure for the purpose of elevating and sustaining the highest level of success in your profession and in your personal life.

Chapter 3 Motivate Team Members: Explained how to create a *Motivation Environment* for each of your team members. When you create a *Motivational Environment*, your team successes will super exceed team cultures absent of motivational environment.

E Pluribus Unum "Out of Many, One"

*Leadership understanding the application of chapters 1, 2, and 3, are prerequisite for the Science and Art of Mental Skills **Part II: Accomplishing Your Visionary Goal***

AN **INDIVIDUAL** WITH A **VISIONARY GOAL** THAT HAS THE ABILITY TO **MOTIVATE TEAM MEMBERS** TOWARDS ACCOMPLISHING **YOUR VISIONARY GOAL**

PART II

SCIENCE AND ART OF MENTAL SKILLS

◆ ◆ ◆

Goals Imagery Relaxation Refocus

Team Members that are High Achievers are Skilled at using Mental Skills.
SHANE MURPHY, PhD
World Renowned Psychologist

PART II: ACCOMPLISHING YOUR VISIONARY GOAL WITH MENTAL SKILLS

Chapter 4: *Goal* setting techniques are the foundation of any mental skills program. Science and Art of Goals is a prerequisite for the remaining mental skills.

Addendum: *Human Brain*-A general discussion of the *human brain* enables a better understanding of the mental skills: Imagery, Relaxation and Refocus. The major purpose of Imagery, Relaxation and Refocus is to enhance the success of all goals within your organizational structure and in your personal life.

Chapter 5: *Imagery* a mental image imitating reality which serves a purpose. One of the major benefits of imagery is building self-confidence.

Chapter 6: *Relaxation* the ability to control stress level; attempting to achieve the *Alpha Stage* for maximum performance.

Chapter 7: *Refocus* mentally achieving focus after stimuli interrupts original focus: *The number 1 cause of failure in profession and life*: **Inability to Refocus**.

"Team members expect you their leader to have a vision and strategy to pursue excellence in achieving the future. They expect you to have a vision and a strategic plan for taking them to a new success." (Sullivan & Harper, 2005, p. 235) When your leadership implements mental skills you are allowing each team member to explore their unlimited boundaries of full potential. (Lynch & Chungliang, 2006)

Part II presents a successful strategy for *Accomplishing Your Visionary Goal* and elevates your leadership to its highest sustainable level, without destroying the rest of your life.

All roads to successful Accomplishment of Your Visionary Goal are through the mind. All successes are first born and won there and all failures are first born and lost there.
WEST POINT MILITARY ACADEMY

Leadership and Mental Skills

Many leaders spend casual or no time preparing team members with the application of mental skill techniques. Leaders who implement physical and technical training with the exclusion of mental skills is analogous to a basketball program implementing defense and offense training with the exclusion of transition.

Leaders who accomplish sustainable success know the secret: empowering team members with mental skill techniques. Mental skills are an interdependent cognitive process in which every detail may be properly addressed.

Each mental skill is presented allowing you to pick and choose the glove of best fit for you and your team members. Mental skills of Goals, Imagery, Relaxation, and Refocus are presented in two major categories:

1. SCIENCE: Knowledge of the Skill   **2. ART:** Application of the Knowledge

Understanding science and art of each skill enables elevating leadership to its highest sustainable level. You will discover throughout the following chapters each of these psychological skills is interrelated and are an enormous benefit in your personal life and your team member's personal life. Mental skills are not only professional skills; just as importantly, they are life skills.

Scientific Research: Important Disclosure

On several occasions throughout the remaining chapters scientific research will be reported. It is important to note that scientific research does not prove anything; never has and never will. Scientific research can only indicate no significant difference or a significant difference at a certain level. The significant difference may be a negative or a positive result. In order for scientific research to be robust it must be repeated with like circumstances by other researchers with similar results.

Bias Research: Involves personal agenda affecting research result: Following are two examples.

- Global Warming: Often those with a bias agenda use scientific research to falsely prove their biased agenda. In global warming research, some scientists manipulated data to fit their personal agenda. Remember: "Figures don't lie but a lot of liars figure"
- Super Bowl: Another example is that of a major network who deliberately misled their audience to increase viewer ratings. In this case the network reported research conducted during the Super Bowl proved domestic violence increased. This research was conducted but the results were: there is a difference in domestic violence during the Super Bowl, but that difference is not significant. Meaning the results probably are due to how the research was conducted or some other variable.

JUMP FLY JUMP

Scientific researchers themselves have been known to misinterpret results. Such is the case of the scientist who researched the fly. First, he placed the fly in a glass container and yelled "Jump". The fly jumped and he measured the jump. Next, he removed one leg of the fly and yelled "Jump." Measured the jump and it was less than the prior jump. He repeated this until he removed all six legs from the fly and yelled "Jump." The fly didn't move. He repeated "Jump." Once again the fly didn't move. This research scientist made the following conclusion: When you remove all the legs from a fly it becomes deaf.

We hope this didn't happen but it does illustrate one should understand how others and even some researchers may incorrectly interrupt the results of scientific research.

***Caution:** Two major types of research.

1. Academic/Scientific Research: No agenda and must expose limitations and biases. Academic research is carefully evaluated for biases and limitations through several layers of expert committees.

2. Nonacademic Research: Often has an agenda and is under no obligation to expose bias and funding.

CHAPTER 4:
GOALS

◆ ◆ ◆

INTRODUCTION

AN **INDIVIDUAL** WITH A **VISIONARY GOAL** THAT HAS THE ABILITY TO **MOTIVATE TEAM MEMBERS** TOWARDS ACCOMPLISHING **YOUR VISIONARY GOAL**

Accomplishing Your Visionary Goal with Team Member Goals

Goals   Imagery   Relaxation   Refocus

> *When you go out and achieve your goal that is something no one can take away from you.*
> CARL LEWIS

Olympic Gold Medalist

Developing the mental skill of *Goals* empowers leadership and team members to successfully pursue the future. *Goal Setting* may be applied in one of two ways: the wrong way or the right way.

The proper order of the four *goals* within your organizational structure is vital. Chapter 4 explains the proper order: the knowledge (Science) and the application (Art) of *goals* for successfully pursuing the most productive results towards *accomplishing your visionary goal.*

CHAPTER 4 GOALS: MAJOR TOPICS

Review of Leadership Goals (2)

Introduce Team Member Goals (2)

Science and Art of: Team Member Destination Goals

Science and Art of: Team Member Performance Goals

Art: Implementing Your Team Member Goals Program

Chapter 4 Glossary: Art, Effective Communication, Feedback, Flow, Leadership's Performance Goal, Leadership's Visionary Goal, Motivation, PED, Science, Stair Step Method, Team Member's Destination Goal, Team Member's Performance Goal

GOALS

TEAM MEMBER GOALS

AN **INDIVIDUAL** WITH A **VISIONARY GOAL** THAT HAS THE ABILITY TO **MOTIVATE TEAM MEMBERS** TOWARDS ACCOMPLISHING **YOUR VISIONARY GOAL**

Accomplishing Your Visionary Goal with Team Member Goals

Jackie Joyner Kersey one of the greatest women athletes (8 Olympic Gold Medals) based each of her highly successful and sustainable performances on proper goal setting techniques.

Pursuing your leadership to its highest level is a journey. "On any journey one needs a clear map, a sound vehicle, and sufficient fuel." (Millman, 1979, p. xiii) *Science* of the mind is your map, your body the vehicle; *The Art* of applying the mental skills of goals, imagery, relaxation, and refocusing the fuel.

REVIEW OF YOUR LEADERSHIP GOALS: (2)

1. Leadership's Visionary Goal: Your vision for a targeted outcome within a specified time

2. Leadership's Performance Goal: Your strategy for the *accomplishment of your visionary goal*

INTRODUCING TEAM MEMBER GOALS: (2)

1. Team Member's Destination Goal: Team member's targeted outcome of performance, within a specified time, which relates to the successful *accomplishment of your visionary goal*

2. Team Member's Performance Goal: Team member's strategy, within positional boundary, for successful pursuit of performance.
 *Team Member Goals within your organizational structure are designed for the ultimate purpose: ACCOMPLISHING YOUR VISIONARY GOAL
 Individual team member goals should not be to surpass others, but to constantly surpass their own previous achievements. Individual team members should understand they are not competing with others, but with their own strengths and weaknesses. Whatever a team member has achieved in pursuit of their goals, they should focus on going beyond.

Crucial Responsibility of Your Leadership: Before your team members learn and practice Destination and Performance Goals your Leadership must:

1. Clearly and effectively provide team members with <u>Your Written Visionary Goal</u>. Providing team members <u>Your Performance Goal</u> is optional; however, in most circumstances it is preferred.

2. Your leadership needs to clearly and effectively provide each team member their Positional Boundary and your Expectations within their positional boundary.

SCIENCE & ART: TEAM MEMBER DESTINATION GOALS

Definition   Why   When   How

Definition Team Member *Destination Goal*: What your individual team member hopes to accomplish within their positional boundary over a set time frame, which relates to the successful *accomplishment of your visionary goal*.

SCIENCE: KNOWLEDGE OF DESTINATION GOALS

Why: Primary benefits of a Team Member *Destination Goal* (long term)

Enables team member *empowerment*

Enables team member understanding their piece of *Your Visionary Goal*

Enables *direction* over time frame (a personal road map)

Enables team member setting *Performance Goals*

Enables team member understanding your *expectations within their positional boundary*

When: To Set Team Member Destination Goal

Military: Prior to Mission(s)

Business: Prior to first quarter

Education: Prior to school year

Sport: Prior to season, tournament, special event

Politics: Prior to a campaign

Life: Prior to a certain month, year, season, personal life event, beginning of a life crisis

ART: APPLICATION OF TEAM MEMBER DESTINATION GOALS

How: To Set Team Member *Destination Goal*

A team member develops a written general description of what they expect to accomplish within their positional boundary over a certain time frame, which relates to the successful accomplishment of your visionary goal. *Destination goal* is a long term goal.

Examples: (proper) Team Member *Destination Goal* in different professions

Military: 100% dedication to assigned mission(s)

Business: Focus on positional assignment this quarter

Education: Improve teaching skills this school year

Sport: Dedicate myself to hard work in daily practice throughout the season

Politics: Staying connected with my constituents needs throughout my term

Life: Improve a special relationship this summer

It is leadership's responsibility to educate team members on *How* to develop *Destination Goals* with maximum controllability over its success.

Example: Applied Destination Goal

A 107 member Track and Field team had two meets left in the season. One hundred four team members had reached or surpassed their *destination goal*, which were set prior to the season.

Remember, goals may be flexible and or changeable. In the process of counseling the three remaining team members they adjusted their *destination goal* downward for the remaining meets. The next track meet two of three track members had surpassed not only their adjusted goal, but also their original *destination goal*.

The one remaining team member who had not yet achieved the *destination goal* had a farming accident in which he broke his leg prior to the last two meets. He adjusted his *destination goal* as follows: For the remainder of the season I want be in uniform and support my team members to the best of my ability. He did so, on crutches, wearing his team jersey.

This is one of the benefits as to why goals should be flexible and changeable. The remaining three team members were experiencing too much stress (Chapter 6 Relaxation). Adjusting their goal (flexibility) lowered their stress level for successful accomplishment. Flexibility allows controllability of goals.

Three Criteria your leadership provides before team member sets *Destination Goal*:

- Duplicate picture of your *visionary goal*
- Understand their *positional boundary*
- Understand their *positional expectations*

Stair Step Method: The Best Strategy for developing team member *Destination Goal*

Linear Strategy of *Destination Goal:*

* Where are you now?
* Where do you want to go? (*Destination Goal*)
* What is your general strategy to get there?

FIRST develop Stair Step 1: Record where you believe you are presently in your positional assignment.

SECOND develop Stair Step 3: Record what you would like to accomplish within a certain time frame (*Destination Goal*).

LAST develop Stair Step 2: Record what you believe you need to do (strategy) to accomplish your *Destination Goal*.

Stair Step 3 (second)_____

Stair Step 2 (last)_____

Stair Step 1 (first) _____

*This Stair Step Method is analogous to a pilot's flight plan or planning a road trip.

We believe the Stair Step Method is the most successful strategy presently known for proper development of team member's *Destination Goal*. This method is used at West Point Military Academy.

SCIENCE & ART: TEAM MEMBER PERFORMANCE GOALS

Definition   Why   When   How

SCIENCE: KNOWLEDGE OF PERFORMANCE GOALS

Definition: Team Member *Performance Goal* (short term): Strategy developed within positional boundary by the individual team member for the successful pursuit of an approaching performance.

Individual team member should have ownership and controllability over degree of success for their *performance goal*. Increased ownership equates to increased motivation and assists in the development of team culture. Furthermore, controllability increases self-confidence and empowerment.

Research: U. S. Military Goals (Secret Project)

In the 1970's the United States Military branches combined their efforts in a secret mental skills research program. The results were astonishing. The reviews of this research by the top brass of each military branch discovered two major results:
1. When the guidelines are followed *performance goals* are fail proof.
2. This *performance goal setting program* will change the way we train our soldiers.

Question: The United States Military was not the first nation to train pilots with *performance goal* techniques. Which country do you think was first?

France Spain Italy China Israel Germany Australia Great Britain

Answer: Israel

Why: Team Members Set *Performance Goals* (Short Term)

In almost every profession development of physical and technical skills is necessary. When team members ascend to a higher level of performance, more often than not physical and technical skills have parity when compared to other like performers. The difference in who accomplishes the greatest success is in the mind. Team members who develop mental skills will <u>rise to a higher level than</u> those who do not. This is one of the major benefits as to why *performance goals* are important.

Question: Can I Accomplish My Performance Goal without Mental Skills? Yes But!!

On occasion I have been asked, and it is a fair question, "Why do I need mental skills to achieve success?" Each time yielding to the explanation of a West Point cadet who was an avid practitioner of mental skills.

I asked the cadet, "Why are mental skills important to you?" Here is the analysis he presented.

"On rare occasion we are afforded the opportunity of going to New York City on the weekend. We can travel there in a Lamborghini or a rust bucket. Either way, we will get to New York City and achieve our goals. However, traveling in a Lamborghini we will achieve more success in a shorter time than if we used a beat up old vehicle. The Lamborghini is comparable to using mental skills in pursuing success."

Continuing to pick his brain, "I assume when you refer to achieving success in New York City you are referring to R and R (rest and relaxation)?"

In his reply this young cadet almost started to smile (kinesis). However, he quickly recovered (refocused) and replied, "Yes sir; that is one way of stating it."

The thought raced through my mind to ask if he had ever seen a *purple sweater* during his R and R in New York, but I quickly refocused and kept this thought to myself. (Chapter 5 Imagery "The Purple Sweater")

Can you achieve goals successfully without mental skills? Yes. So the question becomes, how does one want to travel in pursuit of a *performance goal*, in a rust bucket or a Lamborghini?

Five Major Benefits of Team Member *Performance Goals*:

1. Goals Improve Self-Confidence
2. Goals Improve Intrinsic Motivational Environment
3. Goals Identify Performance Strengths and Weaknesses
4. Goals Improve and Accelerate Successful Performance
5. Goals Improve Leadership and Team Members Effective Communication

1. TEAM MEMBER *PERFORMANCE GOALS*: IMPROVE SELF-CONFIDENCE

Team members with self-confidence do not fear failure in pursuit of a *performance goal*. One of three tenets all great leaders have developed: They do not fear failure. In every instance a more self-confident team member equates to a more competent team member. The same is also true for your team position leaders.

Self-Confidence: A team member's belief of success in accomplishing a *performance goal*.

Lack of Self-Confidence: A team member's belief they do not possess the mental, physical, or technical ability necessary to successfully accomplish a *performance goal.*

Over-Confident: An individual's belief they possess the ability to accomplish an unrealistic *performance goal.*

DISTINCT MENTAL AREAS OF CONFIDENCES

- **Lack of Confidence**

- **Self-Confidence (Flow)**

- **Over-confident**

Notice: In the area of *self-confidence* the State of *Flow* has the greatest possibility to be experienced. Flow may also be referred to as the Zone or Rhythm.

Flow Defined: A euphoric mental state of an effortless, highly successful performance.

To this date, science has been unable to determine exactly how the mental state of flow is accomplished. However, science is getting closer.

Chapter 6 Relaxation discusses the necessity of the brain's Alpha waves being present to achieve flow. The state of Flow is elusive and seldom experienced.

When Flow occurs it is experienced during self-confidence. The possibility of experiencing flow occurs in any profession from leadership to team members. Flow may be experienced during any performance such as a presentation, an interview, a negotiation, a military operation, a sport event, personal life events, and other such circumstances. Individuals I have been associated with in experiencing the state of Flow have been unable to properly describe this euphoric sensation.

If you have experienced flow during performance, can you adequately describe this euphoric phenomenon? It may be comparable to seeing the Grand Canyon, Victoria Falls, or the Great Barrier Reef; then trying to describe it to one who has never been there.

Developing the proper level of self-confidence in yourself and in your team members is a tremendous benefit for *accomplishing your visionary goal*. Self-confidence is the foundation for success. Proper systematic Team Member *Performance Goals* develop self-confidence and empower your team members.

Chapter 1 addressed insecure leaders and the destruction which occurs under this type of leadership. Fundamentally, insecurity is fear. How does one transform insecurity/fear into self-confidence? A systematic *performance goal* program increases self-confidence. Eventually, it will control insecurity and transform insecurity into a confident leader. This accomplishment requires dedicated self-discipline over a frame of time. Anything worth accomplishing can be done so with self-discipline.

2. TEAM MEMBER *PERFORMANCE GOALS*: IMPROVE INTRINSIC MOTIVATIONAL ENVIRONMENT

Motivation Defined: A team member's mental intensity toward successful accomplishment of a performance.

Building self-confidence through effective goal setting is a necessary precursor to motivation. Motivation occurs when a team member has been successful or believes they are about to experience a successful performance. (Discussed in Chapter 3) When your leadership places the correct team member in the proper position and effectively communicates expectations within positional boundary, you have provided a motivational environment. *Performance goals* provide team members and team with purpose, ownership, and direction.

Applied Written *Performance Goal*: Intrinsic vs Extrinsic Motivation

About two thirds through a 16 team track and field invitational, an opposing coach in wonderment asked the following question. "How do you get your kids to do that? When my kids aren't in first, they let up when they know they cannot win. Your kids stay motivated throughout the entire race regardless of their position. You have such happy kids. They are so nice, and then they go out and proceed to kick everybody's butt. How do you do that?"

Before I could reply, one of our young female thinclads excitedly interrupted our conversation. With a big smile on her face, and her goals journal in her hand she reported, "Coach, I have surpassed two of my goals, tied one, and have one more event." I accepted her journal and gave her verbal and recorded feedback. She had not finished first in her three events. This conversation demonstrates written goals develop **intrinsic motivation**, whereas, the opposing coach's team members were relying upon **extrinsic motivation**.

As soon as the young female runner departed, the opposing coach continued, "That's what I mean; how do you keep them motivated throughout their event?"

"Our team members do not compete against another team, or an opposing runner, we compete against ourselves (**Intrinsic Motivation**). Our kids record in their performance journals what they would like to accomplish **(performance goals)** prior to the meet. Their motivation and focus is to reach or surpass their past *performance goals*."

The following day, I received a phone call from their coach asking help for his kids concerning *performance goals* and help setting up a Mental Skills Program. I agreed and did so, providing each of his student/athletes and his coaching staff with *performance goal* journals. Goal Journals improve a closer relationship between leadership and team member. In addition, goal journals identify team member strengths and weaknesses.

***Important:** It is paramount your leadership identifies individual team members needs, then provides the necessary tools (physical, technical, mental) for your team member to properly address and fulfill those needs. Identifying and addressing team member needs is vital in providing a successful motivational environment.

An effective method for discovery of your team members' needs is to develop a one-on-one interview. Developing your interview is fully addressed in the *How* section of this chapter.

3. TEAM MEMBER *PERFORMANCE GOALS*: IDENTIFY PERFORMANCE STRENGTHS AND WEAKNESSES

Every team member has strengths and weaknesses. Written goals present the opportunity for this discovery. Leadership combined with team member's analysis of performance is greatly beneficial in staying on the right track for improving strengths and eliminating weaknesses. When a team member identifies performance strengths and weaknesses, your team member is empowered with purpose, ownership, and direction in pursuing with confidence the achievement of their *performance goal*.

4. TEAM MEMBER *PERFORMANCE GOALS:* IMPROVE AND ACCELERATE SUCCESSFUL PERFORMANCE

The five major benefits of *performance goal*s are interrelated. In the process of explaining this mental skills program to a Division I coaching staff, one of the coaches asked the following question, "How much will these mental skills improve the performance of each player" (team member)? When my reply was a very conservative estimate of one percent or more, the coach answered, "That's all?"

"How many players do you think will participate in tomorrow's game?"

"About 35 or 40"

I followed up his concern in the following manner: "Let's see; one percent times 35 equals 35 percent. You have now increased your team's potential performance success by at least 35 percent." The coach understood and accepted this analogy, which is proper for all professions involving performance.

Developing greater self-confidence and motivation will result in improved and accelerated successful performance. In fact, when you the leader develop a team member with self-confidence and motivation … watch out, you have created a team member who is invaluable in pursuit of your visionary goal. That is unless you are an insecure leader, in which case your team member should … watch out!

It is important you the leader address your own strengths and weaknesses with proper goals. When your team members discover you too are practicing goal setting, this is a major positive in their efforts to realize the benefits of goals.

5. TEAM MEMBER *PERFORMANCE GOALS:* IMPROVE LEADERSHIP AND TEAM MEMBERS EFFECTIVE COMMUNICATION

Research indicates **ineffective communication** in leadership results in 97% of issues and conflicts within organizational structure. Written *performance goals* permit you and your team member to be on the same page. When everyone is on the same page, the result is a greatly improved team performance.

Effective Communication Defined: When you the sender create a duplicate picture in the mind of the receiver.

*Regardless of age, gender, or intelligence level, some of your team members will not have the necessary communication skills required to address certain issues. Leadership feedback is often a communication tool enabling a team member with an avenue to address concerns they otherwise would be unable to properly address. Leadership feedback is critical in effective communication and performance goal setting. Proper leadership feedback will be presented in the HOW section of this chapter.

When: Team Members Set *Performance Goals* (Short Term)

Team Member *Performance Goals* may be developed just prior to performance.

Profession	When
Military	Prior to any desired performance
Business	Prior to any desired performance
Education	Prior to any desired performance
Sport	Prior to any desired performance
Politics	Prior to any desired performance
Life	Anytime

Generic Examples: Team Member *Destination Goal* with compatible *Performance Goal*

Military:

(Destination Goal) 100% dedication to this mission
(Performance Goal) 100% dedication to my training of physical, technical, and mental skills

Business:

(Destination Goal) Focus on positional assignment this quarter
(Performance Goal) Develop a strategy for achieving my positional expectations

Education:

(Destination Goal) Improve teaching skills this school year
(Performance Goal) Enroll in a college class to improve a weakness

Sport:

(Destination Goal) Dedicate myself to consistent hard work throughout the season
(Performance Goal) Focus in practice on improving my physical, mental and technical skill

Politics:

(Destination Goal) Staying connected with my constituents needs throughout my term
(Performance Goal) Focus each day on constituents needs with verbal, written and technological communication

Life:

(Destination Goal) Improve my relationship with a loved one this summer
(Performance Goal) Focus on becoming more understanding and compassionate

ART: APPLICATION OF TEAM MEMBER PERFORMANCE GOALS

How: Team Members Set *Performance Goals* (Short Term)

Team Members Set (Fail Proof) *Performance Goals*: Two Structures

- Technical

- Mental

1. **TECHNICAL STRUCTURE:** Team Member *Performance Goal* (3) Criteria

    - **Written:** *Performance Goals* should be Written in a journal or diary. Journal requires leadership feedback. In your personal life, record in a diary (YEO, Your Eyes Only)
    - **Measurable:** *Performance Goals* should be Measurable (0-9 chose any 2 consecutive numbers)
    - **Feedback:** *Performance Goals* should include leadership Feedback (verbal/coded) as soon as possible after performance

    o **Written:** Research indicates Written Goals are 40 to 80% more achievable when compared to unwritten goals.

Performance goal setting is not an event; it is a disciplined process. Therefore, it is very beneficial to record performance goals in a journal to evaluate progress over time and to assist in development of one's future performance goals. In developing the next performance goal a good baseline is reviewing the past one, two, or three performance goals with the recorded feedback in their journal.

Retail stores have blank journals at a minimal cost. I highly recommend this in the beginning when team members are practicing *performance goal setting*. Once team members are proficient in setting goals you may wish to continue with the same journal. Should you choose to purchase professional goals journals, Contact: leadershipmentalskills@gmail.com

Important: Team member's *written* systematic *performance goals* are one of the important building blocks for the foundation of your Mental Skills Leadership Program. The benefits are enormous in pursuit of excellence throughout your organizational structure.

APPLIED *PERFORMANCE GOAL*

Business

Several decades ago one of the three major American car companies was in danger of bankruptcy. This company brought in a new CEO who had been highly successful with another of the Big 3 American car companies.

One of his first decisions was to assemble all his position leaders and have each write down what they would like to accomplish within a given time frame. At the end of this time he again assembled the same team members and each evaluated how well they believed their *performance goal* was attained. Team member *performance goals* were a major factor in the new success. This CEO turned around the nearly bankrupt company and made it successful. Mental Skills Leadership applies to all professions and life.

o **Measurable**: Team member Goals should be Measurable

Prior to performance each team member records their written *performance goal*(s), including their strategy for accomplishment, in their personal journal. Directly beneath the written *performance goal* in their journal, the 0-9 scale should be available. After performance your team member will circle any two consecutive numbers. (0 1 2 3 4 5 6 7 8 9).When your team member completes performance they should, as soon as possible, *measure* their performance on the 0-9 scale, then seek leadership feedback.

It is important your team member record how they felt concerning pursuit of achievement in their *performance goal*. All goals in all professions may be *measured* with the scale 0 1 2 3 4 5 6 7 8 9. This is a common scale used in psychology, counseling and other scientific/academic research. In the rare event Flow is achieved, 10 may be recorded at the end of the 0-9 scale.

*After your team member concludes performance, leadership should provide feedback; the sooner the more meaningful.

APPLIED *PERFORMANCE GOAL*

Military

Recently, coalition forces captured a major target without causalities. In this military operation a speculative goals journal might look as follow:

Pre Performance

> *Destination Goal*: Successfully capture our target within our specified time limit *Performance Goal*: Focus on my mental and technical training, regardless of circumstances, in accomplishing my assignment

Post Performance Evaluation

> To what degree did I achieve my goals: (team member circle any two consecutive numbers (0-9)
> After their mission if the military operatives were to give themselves a score on *performance goals*, there is no doubt it should be a 10.

For my next performance I believe I need to _____

If the above example were to be used in a journal provide a small space for leadership coded feedback. Example: a small star with number or numbers recorded beside star points in team member journal. (Described in the How section of Feedback)

    o **Feedback:** Team member seeks Leadership Feedback

The enormous benefits of *Leadership Feedback* for team member *performance goals* require this discussion to be presented in the following sections:

Definition   Why   When   How

Definition Feedback: (1 on 1) Leadership's counseling comment on team member performance immediately or soon after performance. (Verbal and Recorded)

Why: Leadership Provides Team Member *Performance Goal Feedback*

Effective systematic *feedback* is a necessity for elevating and sustaining your successful leadership. (Williams, 2009) It is important you understand what to reinforce when communicating *feedback*. It is easy to praise a team member who has just completed a great performance. It is less natural to praise a team member who has tried their best and was not successful. It is very important to your team members that you provide *feedback* for effort as much as you do results. Providing *proper feedback* for effort is *an enabler for motivation and trust between leader and team member*. **Caution:** Never compliment poor effort; in doing so you compromise your leadership's character.

Effective Communication between Leadership and Team Member is vital. Stated earlier, some of your individual team members regardless of age, gender, or intelligence level may lack the proper social skill to professionally express personal concerns and issues. Leadership's verbal and written *feedback* with individual team members will definitely bridge this communication gap.

Verbal and Written *Feedback* is an <u>enabler in keeping quality team members</u> whose motives and behavior might otherwise be misunderstood. *Feedback* assists in limiting or eliminating unnecessary team member dismissal. This leadership technique is not effective with an insecure individual in a leadership position.

Feedback benefits Your Leadership's Ability to keep your finger on the pulse of Your Team Culture. The impact of *feedback* often results in immeasurable positives. Your leadership *feedback* allows team members to know you are personally interested in their *Who Am I* and *What Am I*.

The *Feedback* process allows leadership to monitor proper *Performance Goal Setting* and to observe that their *Performance Goals* are compatible with their Destination Goal. Proper goal setting techniques implemented by team members with leadership *feedback* provides numerous benefits for leadership and team members.

When: Leadership Provides Team Member *Performance Goal Feedback*

It is important your leadership provides *feedback* as soon as possible after performance. When your team member presents the goals journal after their performance, listen and give the team member verbal counseling *feedback*, as mentioned earlier. Plus you may wish to draw a star (top of page) then place a number on one or more of the star points. Provide a small area for this visual *feedback* (star). As you record your visual *feedback*, your team member will definitely study your paralanguage and kinesis during this process.

On the first few occasions, invariably the team member will ask what your code means. "That is my personal code for how well you are progressing." This allows your team member to realize you personally care about them and the importance of how well they are achieving. Performance progress is measured by how well a team member is on track in achieving their destination goal.

How: Leadership Provides Team Member *Performance Goal Feedback*

How your leadership provides *feedback* for individual performance should be kept confidential. *(Leadership/Team Member Relationship; Equivalent to Doctor/Patient Relationship)* Performance *feedback* requires trust and confidentiality between leadership and team member. If trust or confidentially is violated serious damages will result in your team member relationship. Trust may be destroyed in seconds, while building trust takes longevity.

The preceding discussion focused on three important **Technical Structures** of *Team Member Performance Goals*, which included a discussion on *written, measurable,* and *feedback.*

The following discussion focuses on **Mental Structure** of *Team Member Performance Goals.*

2. **MENTAL STRUCTURE:** Team Member *Performance Goals*: (4 Criteria)

- * **Challenging:** *Performance Goals* should be Challenging
- * **Realistic:** *Performance Goals* should be Realistic
- * **Flexible:** *Performance Goals* should be Flexible
- * **Controllable:** *Performance Goals* should be Controllable by team member

Challenging: Team Member *Performance Goals* should be *Challenging.*

Goals which have minimal *challenge* will result in the team member approaching their goal with minimal motivation, resulting in a negative effect on their subsequent goals. Written goals in a journal create a history of past performances, enabling a picture for today and tomorrow's performance goals. Team members should always *challenge* themselves in improving their pervious performances.

Realistic: Team Member *Performance Goals* should be *Realistic.*

Performance goals set beyond your team member's skills will result in too much stress, which decreases motivation and self-confidence for successful achievement. *Realistic* performance goals assist in developing self-confidence and increased motivation. Your leadership counseling assists team members in developing *realistic* goals.

Flexible: Team Member *Performance Goals* should be *Flexible.*

On numerous occasions unexpected circumstances will occur. Such circumstances could affect your *performance goal* as to make it extremely challenging or even unrealistic. Examples of the unexpected: weather, illness, injury, equipment failure, power failure, etc. Usually, the unexpected is in the form of a negative circumstance. However, on occasion the unexpected may be a positive.

MILITARY: FLEXIBLE GOALS A NECESSITY

In a major military theater in the Middle East the Coalition General in command was asked the following question, "How soon do you anticipate having to go to plan B?" The General replied, "After the first shot is fired." Being *flexible* is equivalent to being prepared.

No matter where you go or what you hope to accomplish, sooner or later Murphy's Law will be enacted. Definition of Murphy's Law: Something undesirable bound to happen because it hasn't happen yet. Murphy's Law is another reason for *flexibility*.

SINGLE NUMBER GOAL

Remember: When using numbers in goals give yourself a range. (*Flexibility*) Our defensive back in football, an outstanding athlete, set the following *performance goal*: I want to make three interceptions in today's game. The opposing coach had scouted us and was aware of this young man's exceptional talent, and therefore, not one pass was thrown in his coverage. This fine athlete was disappointed and confused because he didn't reach his *performance goal*.

Now our defensive back was ready to listen to leadership's goals counseling. Next game he designed his *performance goal* as follows with *flexibility:* In today's game I want to be in position to intercept if the game situation permits. Should the opponent not throw in his coverage, he is able to feel great success. His post performance evaluation could be an 8-9. *Being *flexible* in setting goals translates into being prepared. When possible, avoid using numbers or if you find it necessary to use numbers give yourself a range *(flexibility)*.

Controllable: Team member should have *Controllability* over their *Performance Goal's* degree of success.

In developing a *performance goal*, the team member strives to achieve the greatest possible *control* over its success. In earlier discussion of fairness, it was stated there is no such thing as total fairness, however your sincere effort towards fairness validates your character among team members. The same is true of complete *control* in performance goal setting by team members. There is no such thing as complete *control*. It is important your team member designs as much control as possible (*controllability*) into their *performance goal*: more *control*, more ownership; more ownership, more motivation to achieve.

APPLIED PERFORMANCE GOAL LESSON

OLYMPIAN GOLD MEDALISTS

Carl Lewis

Carl Lewis and Jackie Joyner Kersey are a great study of elite performers who valued and religiously practiced performance goal setting. **None** of their super performances **involved performance enhancing drugs (PED)**; instead they used the power of the human brain.

In Carl and Jackie's era many leaders, team members, and sports writers did not fully understand the mental strategy of performance goals. An excellent example of this misunderstanding is found in a major

United States newspaper in 1980s when Carl Lewis finished second in the Olympic 100 meter dash.

As you read this article what do you think? Was Carl Lewis's *performance goal Measurable, Challenging, Realistic, Flexible;* and did he have *Controllability* over his performance goal success?

Lewis's *Performance Goal*: **Run the fastest race I have ever run**

CARL LEWIS REMAINS A RIDDLE

News Commentary

Seoul, South Korea 1988 Summer Olympics- It's as if Carl Lewis were an actor instead of an athlete. As if he had just enjoyed taking a curtain call for a supporting role in a Broadway hit instead of having lost the biggest race of this career in the Olympics.

Upon finishing second to Ben Johnson in the 100 meter dash Carl Lewis remained a riddle wrapped in a red-white-and-blue U.S.A. warm-up suit.

In Los Angeles, he didn't know how to react when he won four Olympic gold medals. Now he didn't know how to react when he didn't win a gold medal.

In his aloof manner, he spoke as unemotionally as a talking computer about how he was "pleased with his performance" of 9.92 seconds although the Canadian set a world record with 9.79.

Pleased? When he didn't win history's best 100 meter race? Performance? When he was a relatively distant second banana?

When the 27 year old American sprinter and long jumper was reminded that most Olympic silver medalists are more disappointed at not having won a gold than they are glad to have finished second, he disagreed.

"The Olympics," he said, "is about performance, the best you can do. I'm pleased I ran the best I could."

Appealing words. But only when uttered by someone who had finished 4th or 8th or 48th, someone on his way up. Not by someone who had hoped to stay up there, someone with an opportunity to be the first to win the Olympic 100 twice, someone who had to win the 100 in order to go for four gold medals again.

But in contrast to Carl Lewis's soft words, maybe Ben Johnson's hard words reflected why he won and why Lewis didn't. "The important thing," Johnson said firmly when about having set the world record, "was to beat Carl Lewis."

That's the way a real competitor should talk. No clichés. No avoiding the issue. No memorized lines. Throughout his comments, Lewis was so cool he was cold. His only words of praise for Johnson were a quick "He ran a great race, he ran a great time."

While Johnson took nearly two hours to reach the interview area, Lewis's only deference to Johnson's victory occurred when he momentarily started to sit down at the gold medalist's microphone, then moved to the silver medalist's microphone. At least outwardly, Lewis never seemed to understand what was at stake.

This wasn't just another Olympic 100-meter dash. This was the Olympics' answer to Ali-Frazier III, to Nicklaus and Palmer in the 1967 U.S. Open, to Borg and McEnroe in the 1981 Wimbledon final.

In his approach to their celebrated confrontation, Lewis talked about how he intended to "focus on just trying to run the best race I can" rather than focusing on beating Johnson. But if Lewis were not focusing on Johnson, why did he turn his head three times during the race to see how far Johnson was ahead of him?

Carl Lewis expected Johnson to burst out of the blocks into the lead. But the American had hoped to accelerate until 70 meters and then maintain that speed while Johnson decelerated.

"It isn't a sense of coming from behind," Lewis said three days earlier. "It's a sense of me continuing the momentum and the other athletes coming back because they focus more on the first half." But Johnson never came back.

As leader you will always have your critics. It is your role and duty to teach position leaders and team members the proper structure of *performance goal* setting. The basic fact is: setting *performance goals* will be more beneficial in achieving your visionary goal than any other method we presently know.

Not end of Carl Lewis Story: Three days later the young athlete from Canada tested positive for steroids and was stripped of his gold medal, which now belongs to Carl Lewis.

I heard Carl Lewis make the following statement, "When you set your goals and you go out and achieve them; that is something no one can take away from you."

*Proper goal setting by team members is the most effectively designed strategy for *accomplishing your visionary goal.*

PERFORMANCE GOAL VS DESTINATION GOAL

Important Lesson from Jules

Our home in Southern Illinois sits on a bluff overlooking the Mississippi river. At the bottom of the bluff, right beside the Mississippi, is a nine mile stretch of secondary road. My wife and I often jog on this seldom traveled route. The wild life such as the bald eagles, deer, wild turkeys, even copperhead and rattlesnakes frequent our exercise route.

The river traffic is also a pleasant attraction. It is a tranquil environment. The Delta Queen, Mississippi Queen, barges with varied cargo, local fishermen, and trains on the track bordering our jogging road were always a welcome sight in this peaceful setting.

Alongside our jogging route sat a modest cabin whose occupants were a little ole man, a retired lawyer whose name was Jules, and his sweet petite wife. Whenever they were outside on their elevated front porch in their rocking chairs, they would insist my wife and I come up to visit and have a cup of tea or coffee.

On one occasion Jules insisted we take a guided tour inside their lovely cabin. Much to our surprise one of the rooms was filled entirely with trophies, big first place trophies. Jules was a world class marksman. The trophies were from all around the world.

I couldn't pass up the opportunity to pick Jules' brain on his secret to such marvelous success. He was eager to respond to my various questions. I asked Jules what made him so successful at this highly completive sport.

"Our shooting matches were always close in scores until the final few rounds. That is when I would excel." When I asked what occurred or what happened, with a smile on his face he replied, "The younger shooters get the lump." He placed his right hand under his chin and around his throat. "They choke coming down the stretch."

Continuing to pick his brain, "Why did they choke when you did not?"

Listen carefully to Jules' response: "At this point many of the others would *get their eye on the prize (Destination Goal) and forget about what it takes to get to the prize.*" (*Performance Goal*)

The lyrics of an old spiritual hymn, "One day at a time sweet Jesus one day at a time" are applicable to *performance goal achievement. One performance at a time or as in Jules case; one shot at a time.*

It is a true example for leaders and team members in knowing why, when, and how to focus on *Performance Goals* and when not to focus on the Destination Goal. Team members focusing on *performance goals* will greatly increase successful accomplishment of their Destination Goal. During performance, focusing on Destination Goal will take away from your Performance goal. "Focusing on outcome will take away from what you are trying to accomplish". (S. Murphy, personal communication, July 1997)

Many times in the profession of sport when teams compete for an International Title or National Championship, you may recall statements such as: This team has an advantage because they have been here before, where as it is the first appearance for their opponent.

The difference: Those who have been there before have learned it is not about focusing on outcome, but rather on personal performance. When focusing on *performance goals* you increase the chances of successful outcome.

> *Team member goals help the individual team member to get out of their own way in pursuing success.*
> T. HUGHES
> Philosopher

ART: IMPLEMENTING YOUR GOAL SETTING PROGRAM

> *The process of slowing down and bringing team members into the conversation about your system and your team culture is one that is not done enough, but nevertheless is critical. Goleman et al. 2002*

The most effective, most productive process any leader in any profession can do is to implement a team member goal setting program.

In the 1970s and 1980 the presentation of mental skills seemed extremely complicated. Today science has evolved the presentation of mental skills with a simple, clear, and effective implementation. It is now possible to use the KISS Method in HOW to implement the mental skills program of *performance goals*.

It is greatly beneficial for you and your position leaders to practice written goal setting to enable clarity of implementation to your team members. Should you decide to start small you should focus on those team members with the greatest potential and or the most important positions.

Your educational structure should be developed with the trickledown theory. It is important your positional leaders have belief, conviction, and passion concerning the development of your mental skills program. Having done so, these qualities will permeate throughout your team culture.

*Teaching, Counseling, and Coaching are beneficial communication skills for successful implementation of your goals program. Should you choose to purchase professional Mental Skills Journals for your Goals Program contact: leadershipmentalskills@gmail.com

FOUR TEAM MEETINGS

Implementing Your Goals Program

*If implementing your goals program is not in your immediate future, you may wish to lightly skim the listings under Team Meetings 1, 2, and 3. However, it is important to read in depth starting at Team Meeting 4.

Meeting 1 Educate: Overview of Mental Skills

- Hand out and Introduce Mental Skills Journal (MSJ)
- In the MSJ on first page provide space for *Your Visionary Goal*
- Have each team member record *Your Leadership Visionary Goal* from a visual aid of your choosing (handout, overhead projector, power point and etc.)
- Team members record *Your Leadership Performance Goal* in MSL journal *(optional)*
- If you have purchased professional goals journals: Introduce each page of the journal and give a brief overview of each mental skill as presented in the journal.
- List each team position, position boundaries, and position expectations in the appropriate space of the goals journal

Needed Materials:

Mental Skills Journal with pen or pencil

Visual Aids: Chalkboard, Smart Board, Overhead, Power Point

Optional: Handouts of your choosing

Optional suggested Film: Top Gun, Cool Runnings Team meeting #1 or 2

Time allowed for questions and discussions

*When team members become verbally interactive it is beneficial for everyone in your program

Meeting 2: Practice Team Member Destination Goal

- Leadership reiterates Team Positions and Expectations within each Team Position
- Team Member record in MSL journal Team Positions and Expectations within each Position
- Explain how team member's *Destination Goal* is to be designed for the successful accomplishment of Leader's Visionary Goal
- Stair Step Method, explain in detail; everyone should understand this method

Stair Step Method: The Best Strategy for **Team Member Destination Goal**

Stair Step 1: FIRST- Record where you believe you are presently in your positional assignment. Stair Step 3: NEXT-*Destination Goal*; Record what you would like to accomplish within a certain time frame. Stair Step 2: LAST-Record what you believe you need to do to accomplish your *Destination Goal*.

                        Stair Step 3 (second)_____

           Stair Step 2 (last)_____

Stair Step 1(first)_____

This stair step method is analogous to a pilot's flight plan.

Needed Materials:

    Mental Skills Journal with pen or pencil
    Visual Aids: Chalkboard, Smart Board, Overhead, Power Point
    Optional: Handouts of your choosing
    Optional: Related film of your choice

Time allowed for questions and interactive discussions
Practice written Destination Goals with provided sheets in back of Mental Skills Journal
One on one Leadership or Position Leader's counseling team members as needed

Meeting 3: Implement Team Member Performance Goals

- Introduce *Performance Goals*
- Educate how to set *Performance Goals*
- Practice hypothetical situations
- Design personal *Performance Goals* within positional boundary

Needed Materials:

Mental Skills Journal with pen or pencil
Visual Aids: Chalkboard, Smart Board, Overhead, Power Point
Leadership explains how team member *performance goal* relates to team member destination goal
Handout: 8 lane *performance goal* exercise

8-Lane Performance Goal Exercise: It is preferable to name a competition which your team members are unfamiliar. In this example we'll use swimming. However through magic you make each team member a world class swimmer. In each lane an Olympic competitor is presented.

100 Meter Freestyle 8 Lanes

Lane 1: France_____

Lane 2: Canada_____

Lane 3: United State_____

Lane 4: Spain_____

Lane 5: Italy_____

Lane 6: Each team member writes in their name_____

Lane 7: United Kingdom_____

Lane 8: Germany_____

Have each team member circle the lane number in which they perceive to be their greatest challenge in swimming the 100 Meter Freestyle. Having done so, provide the correct answer: Lane 6; the greatest challenge is within yourself. This exercise is analogous for how to set *performance goals*. One's greatest challenge comes from within. You may wish to alter this exercise with choosing competitor names of your choice; such as naming position leaders in different lanes.

In addition, allow time for questions and discussions; interactive discussion is a plus. Practice written *Performance Goals with provided sheets or in back of MSJ. Leadership may need to provide one-on-one counseling as team members practice written goals.*

*It is possible an additional team meeting may be necessary for *Performance Goals*

Meeting 4: Interview

Important: This interview will provide numerous unexpected discoveries. Leadership/ Team Member Interview is invaluable

- Written interview should be provided in Mental Skills Journal.
- Leadership conducts a personal interview with each team member.
- Leader provides team member with identical written copy of interview questions.

- Written copy enables leader and team member to be on same page and on same item of discussion during the interview process. Provide space for note taking.

The following are sample questions for team member interview. If someone other than leadership (you or your position leaders) conducts this interview the possibility of <u>disconnect</u> between leadership and team member is greatly <u>increased.</u>

SUGGESTED QUESTIONS FOR YOUR INTERVIEW

The science of human behavior is a major factor in development of this interview. After reading the interview questions, you will bring clarity of meaning if you first interview yourself.

Interview question 1: If you could have one wish for yourself what would it be?_____ Rational (why)_____

_____

Interview question 2: If you had one wish for the world what would it be?_____ Rational (why)_____

Interview question 3: Review this team member's position, position boundaries, and position expectations. Why have you sought this position?_____
Rational (why)_____

Interview question 4: To what degree do you understand your positional boundaries? (team member circle any 2 consecutive numbers)

0   1   2   3   4   5   6   7   8   9

Interview question 5: To what degree do you understand your positional expectations? (team member circle any 2 consecutive numbers)

0   1   2   3   4   5   6   7   8   9

Interview question 6: What do you believe to be a personal strength in your positional assignment?_____

Rate your mentioned strength-(circle any 2 consecutive numbers)

0   1   2   3   4   5   6   7   8   9

Interview question 7: What do you believe to be a personal weakness in your positional assignment?_____

Rate your mentioned weakness-(circle any 2 consecutive numbers)

0   1   2   3   4   5   6   7   8   9

Interview question 8: Is there anything I can do to help you enjoy and best perform your positional assignment?_____

Optional: The following space is provided for your thoughts in developing additional interview questions for your team members.

Leadership's practice Interview question 1: Design this question concerning team members personal opinion not related to team position

_____

Leadership's practice Interview question 2: Design this question concerning team member's position _____

Leadership's practice Interview question 3: Design this question asking team member if they have any personal or professional concerns.

_____

*This interview is designed to increase the possibility of getting inside the head of your team member. Your Leadership/Team Member Interview is invaluable for developing your team culture and for accomplishing your visionary goal. At leadership's discretion periodically call team meetings.

SIX MAJOR PURPOSES FOR PERIODIC TEAM MEETINGS

Attending to the following six basic purposes of those you serve is crucial for sustainable success of accomplishing your visionary goal.
Six Major Purposes for Periodic Team Meetings:

1. Allows your leadership to keep your finger on the pulse of the team and team member progress

2. Reinforces leadership's commitment to the visionary goal which allows team members an uninterrupted motivational environment

3. Becomes an advantage for your leadership in effective communication with your team members

4. Allows team members to know you really care

5. Allows team members to know they are an important part of the team culture

6. Assists in minimizing and controlling satisfaction and complacency

Leaders who understand the science of human behavior have a distinct advantage in elevating leadership to its highest sustainable level. It is paramount for your leadership to understand the basic needs of your team members.

Basic Needs in Life: (3) In our personal life and in our marital relationship there are three basic needs requiring nourishment and cultivation:

- * Physical Needs: Monetary, Shelter, Food
- * Emotional Needs: A consciously nourished cognitive connection between two committed individuals
- * Sexual Needs: Fulfilling ones normal sexual desires

*Understanding basic needs enables your leadership insight into your team member's behavior and performance quality.

BASIC NEEDS IN HUMAN BEHAVIOR

Understanding the needs of your team members is an essential function of your leadership. Team members *performance goals* with leadership *feedback* will enable you to successfully discover, understand, and address team member needs.

The number one drive (need) in the human being is the drive for survival. The number two drive (need) the sex drive. Keep in mind we are not suggesting you provide either of these needs under your leadership. However, it is important in understanding your team members' behavior and performance.

Recently, a highly successful leader related the following observation: "Leaders, who have their personal lives straightened out, have successful professional performance. Leaders who don't, that eventually reflects in their performance." The balance of *Who Am I* with *What Am I* Chapter 1 included understanding the importance of *What Am I* with *Who Am I*. Goals enable leadership to discover and appreciate the needs of each team member. When your team members believe your leadership sincerely cares about the *Who Am I*, it becomes advantageous in all facets of your leadership sustainable success. Understanding needs of your team members is important and often times comforting and enjoyable.

SCENARIO: STRANDED ON AN ISLAND

Survival or Sex

You are on the largest and best cruise ship ever to set sail. The ship starts to sink. Only two survivors make it to a very small island which is about 20 meters in diameter. You and another gorgeous member of the cruise are stranded on this island for days and food supply has run out. Fortunately, a small sail boat arrives. Unfortunately, the small boat can rescue only one of you. The captain of the sail boat makes you the following proposition. You have two options.

1. You may make love to your gorgeous partner and if you do, you will remain on the island to eventually die. The captain will instead rescue your gorgeous partner.

2. If you chose to leave immediately, the captain will instead leave your gorgeous partner eventually to die.

You will under these circumstances chose option 2, in which you will choose to survive. **The number one need in the human being is the drive for survival.** Maybe the answer to this scenario varies with age or maybe not. What do you think? You are correct; there are exceptions. In some instances the exception is that of a hero.

Other Team Member's Basic Needs

* Need to contribute: Team member voice being heard by leadership
* Need to feel worthy: Leadership values team member performance
* Need to feel ownership: Team member purpose within positional boundary
* Need to feel competent: Team member feeling skillful and proficient in successful accomplishment of positional expectations

Your goals program will successfully address the ever changing personal and professional needs of each team member. Individual team member goals are the foundation of your Mental Skills Program.

The skills of Imagery, Relaxation, and Refocus are beneficial to enhance the success of team member *Performance Goals*. (Addressed in Chapters 5, 6, 7)

*In the final analysis, success or failure of your visionary goal comes back to your team members and how you treat them ... no exceptions.

> AN **INDIVIDUAL** WITH A **VISIONARY GOAL** THAT HAS THE ABILITY TO **MOTIVATE TEAM MEMBERS** TOWARDS ACCOMPLISHING
> # YOUR VISIONARY GOAL

CHAPTER 4 GOALS

SUMMARY

Accomplishing Your Visionary Goal
with Team Member Goals

If you do nothing else, implementing a goal setting program including written goals with feedback will provide numerous desired benefits. Your goals program lays a solid foundation for future development of imagery, relaxation, and refocusing skills.

*Your Leadership understanding team member basic needs is an asset for a highly successful goals program. A successful goal setting program greatly enhances and accelerates the success of *Accomplishing Your Visionary Goal.*

CHAPTER 4 GOALS: MAJOR TOPIC POINTS

Review Leadership Goals (2)

    Visionary Goal
    Performance Goal

Introduce Team Member Goals (2)

    Team Member Destination Goal
    Team Member Performance Goal

SCIENCE AND ART: TEAM MEMBER *DESTINATION GOALS*

Stair Step Method in developing Destination Goals

    Stair Step 1: Record where you are now
    Stair Step 3: Record what you would like to accomplish within a certain frame
    Stair Step 2: Record what you believe you need to do to accomplish Stair Step 3

Leadership Feedback

    Definition Why When How

Goals Benefits (5)

    Goals Improve Self-Confidence
    Goals Improve Intrinsic Motivational Environment
    Goals Identify Performance Strengths and Weaknesses
    Goals Improve and Accelerate Improved Performance
    Goals with Leadership Feedback Improve Leadership and Team Member Effective Communication

Mental Areas of Confidence

    Lack of Confidence
    Self-Confidence
    Over Confident

SCIENCE AND ART: TEAM MEMBER
PERFORMANCE GOALS

Team Member Performance Goal (2) Structures

    Technical Structure:
        Written
        Measurable
        Feedback
    Mental Structure:
        Challenging
        Realistic
        Flexible
        Controllable

Errors in Goal Setting

    Negative terminology
    Another Individual
    Single number

Examples: Destination Goal with compatible Performance Goal in each profession

Applied Performance Goal Lessons

    Carl Lewis
    Jules

ART: IMPLEMENTING YOUR GOALS PROGRAM

Four Team Meetings

    Educate
    Practice
    Implement
    Interview

Basic Human Needs

    Personal
    Professional

Personal Team Member Needs

    Physical
    Emotional
    Sexual

Professional Team Member Needs

    Need to contribute
    Need to feel worthy
    Need to feel ownership
    Need to feel competent

CHAPTER 4 GLOSSARY: GOALS

Art: The application of knowledge

Effective Communication: When you the sender create a duplicate picture in the mind of the receiver

Feedback: Leadership's counseling comment on team member performance immediately or soon after performance. (verbal and recorded)

Flow: A euphoric mental state of an effortless highly successful performance

Leadership's Performance Goal: Your strategy for the accomplishment of your visionary goal

Leadership's Visionary Goal: Your vision for a targeted outcome within a specified time

Motivation: A team member's mental intensity toward successful accomplishment of a performance

PED: Performance Enhancing Drugs

Science: The knowledge of

Stair Step Method: Strategy for developing team member Destination Goal

Team Member's Destination Goal: Team member's targeted outcome within a specified time relating to successful accomplishment of your visionary goal

Team Member's Performance Goal: Team member's strategy within positional boundary for success<u>ful pursuit of performance.</u>

Jackie Joyner Kersey and Carl Lewis, two of the greatest Olympic Track and Field athletes, achieved their greatness with power of the mind. Developing power of the mind is a systematic process with longevity of positive benefits.

In many of today's professions team members may abuse alcohol and drugs causing negative results within your team structure and in team member's personal life. Application of mental skills in your team culture will greatly decrease negative issues for each team member. It is important for your leadership to provide a team environment that includes power of the mind; mental skills. If under your leadership team members were trained in mental skills, your team members would have a choice <u>between positives and negatives</u> in *What Am I* and *Who Am I*.

ADDENDUM: THE HUMAN BRAIN

Relevant Information prior to Imagery Relaxation Refocus

HUMAN BRAIN

What a Brain, What a Brain, What a Brain

A general review of the human brain will enable a better understanding of the brain's involvement for the remaining mental skills *Imagery, Relaxation,* and *Refocus.*
Chapter 5 Imagery, Chapter 6 Relaxation, and Chapter 7 Refocus involves areas of our brain which can be improved with proper exercise. Functions of the human brain are improved with mental exercise such as our physical functions are improved with physical exercise.

Dr. Marilyn Albert, associate professor of psychiatry and neurology Harvard University and Director of Gerontology Research at Massachusetts General Hospital, writes "The good news is this, regardless of age there is not a lot of neuronal (brain cell) loss as we age." To each reader of *Mental Skills Leadership,* regardless of your age, the following cognitive processes are important information as we each age in our professional and personal life.

Dr. Albert's study of 1000 people from age 70 to 80 showed **4 Factors** seen to determine which oldsters maintain their mental agility.

4 FACTORS

Education: Appears to increase the number and strength of connections between brain cells (Remember the suggestion in Chapter 1, enrolling in a college class)

Strenuous Activity: Improves blood flow to the brain

Lung Function: Makes sure the blood is adequately oxygenated

The Feeling: What you do makes a difference in your life

"Mental exercise is important to the brain cells." (Kotulak, 1997, p.162) The cognitive skills in the following chapters of *Imagery, Relaxation,* and *Refocus* are excellent techniques for not allowing your brain cells to seek early retirement.

Our brain, which weighs about 3 pounds, is a gray mass of jelly-like substance that is 80% water. An interesting fact concerning the brain, it feels no pain. Surgeons can operate on the brain with the patient fully awake. The brain is a powerful force in the development of the human race. Our brain is a very complex organ. Much research has been dedicated to the understanding of its functions. Although scientists have discovered many great findings, it is still a drop in the bucket in relation to the undiscovered workings of our brain. For example, scientist have discovered receptors on the brain cell's surface operate like signal boxes, opening and closing "gates" letting in the chemical energy a cell needs to fire an electrical charge.

Scientists are learning how to control these signal boxes. The possibilities of controlling these brain gates are enormous. What if we could train our brain to open and close brain gates, enabling us to produce a sudden burst of *Alpha waves* upon demand? Is it possible elite performers have already, consciously or subconsciously, developed the ability to open and close brain gates for desired functions? (Kotulak, 1997)

Our complicated and marvelous brain may be thought of as a mass of neurons. The brain and spinal cord compose the central nervous system. "The human brain contains about ten billion nerve cells and ninety billion of the smaller glial cells." (Asimov, 1991, p. 32)

A research scientist described all the nerves in our bodies in the following manner; "If it were possible to tie the nerves together in a long string it would reach to the moon and half way back to the earth. Blood vessels stretched out would cover 60,000 miles, and DNA connected would travel past Pluto." (Michael Goodwin, NY Post 2011)

The brain, the nerves, and spinal cord are analogous to a super highway. This highway has trillions of intersections, (neurons, nerve synapses) sharp curves, and steep hills. Whenever the brain sends a message along this super highway it travels at various speeds, which may reach 170 and 250 mph. This allows us to coordinate our physical and mental processes. Our brain operates on approximately 12 watts of energy. (Asimov, 1991)

Research scientists have divided the main part of our brain (cerebrum) on the basis of what certain parts look like and their respective functions.

Brain divided into Two Hemispheres

1. **The left hemisphere controls the right side of our body.** The left side of our brain has the following dominate functions:

    Language

    Math

    Information

    Logic

2. **The right hemisphere controls the left part of our body.** The right side of our brain has the following dominate functions:

Spatial Abilities

Face Recognition

Music

Imagery

The brain is the organ that permits emotion, thinking, creativity (imagery), movements, dreaming, recall (memory), attention (focus), stress (relaxation) and other human functions.

The outer layer surface of the cerebrum is the cortex. The largest numbers of neurons are located in the brain's cortex. The cortex is divided into *four lobes*: parietal cortex, occipital cortex, temporal cortex, frontal lobe.

Parietal lobe: manages sensation, handwriting, body position and etc.
Occipital lobe: contains the brains visual processing system
Frontal lobe: problem solving, judgment and motor function
Temporal lobe: involves memory, hearing, etc.

Note: A short discussion of the **Brain's Memory** is beneficial for a deeper understanding of the mental skill, Imagery.

Memories are made of this: Memory, a trait making humans unique among the earth's creatures, has long eluded efforts to understand why in some people it can slowly slip away with extended age. Scientists are unlocking its biochemical secrets and enhancing memory in ways that promise to speed learning and halt ageing forgetfulness. (Asimov, 1991) Presently, scientists are developing digital brain implants to reclaim short-term forgetfulness.

The brain imprints memory by altering its wiring and developing new connections between brain cells; also old brain cells are strengthened.

This is true when an individual exercises the memory cells such as practicing the mental skill of Imagery.

Products of exercising the brain are increasing memory and delaying dementia. If we all were to live long enough and not exercise the brain, dementia would eventually become an issue.

Summary of Brain Discussion: It is important to understand our brain already possesses the capabilities necessary for greater development of Imagery, Relaxation, and Refocusing. With the proper mental exercise these functions will improve greatly. Proper exercise for learning mental skills is not a random event, but rather a systematic process. Eventually these improved skills will be available upon on demand.

Exercising the brain at any age will develop rewiring of the brain and develop new connections. Any systematic process requires focused self-discipline. Presently, the limits of the human brain are unknown. Is it possible our brain has no limitations?

CHAPTER 5:

IMAGERY

◆ ◆ ◆

INTRODUCTION

AN **INDIVIDUAL** WITH A **VISIONARY GOAL** THAT HAS THE ABILITY TO **MOTIVATE TEAM MEMBERS** TOWARDS ACCOMPLISHING **YOUR VISIONARY GOAL**

Accomplishing Your Visionary Goal with Team Member Imagery

Goals	**Imagery**	Relaxation	Refocus

Imagination is More Important than Knowledge
ALBERT EINSTEIN

Colonel Lou Csoka, founder of the Human Performance Training Center, West Point Military Academy, made the following statement concerning imagery, "There is a definite mind body connection." West Point has the latest technology, in their Performance Enhancement Center, to verify this mental physical connection. Presently, world renowned sport psychologists Dr. Nate Zinsser, a man of great professional and personal character, heads the Mental Skills Training at West Point.

Occipital Cortex contains the brain's visual processing system. Our every thought produces its own unique chemical composition in the brain. *Imagery* research is growing at a fantastic rate. Developing the mental skill of *imagery* is enjoyable and exciting. *Imagery* skills are the most powerful skills you can learn in pursuit of performance success.

Many leaders and team members of both genders in all professions participate in the game of Golf. The sport of Golf requires mental skills more so than any other performance in the Western World. Elite golfers use *imagery*. Because the profession of sport has immediate transparency; on numerous occasions media reports successful performance that involves *imagery*. However, highly successful leaders and elite team members in all profession have developed *imagery* to enhance performance.

CHAPTER 5 IMAGERY: MAJOR TOPICS

Science: Knowledge of Imagery

The Purple Sweater
Scientific Theories

Art: Application of Imagery

Do
Don't

Chapter 5 Glossary: Creative Imagery, Imagery, Psychoneuromuscular Theory, Recall Imagery, Symbolic Learning Theory

Mental Skills of Imagery, Relaxation, Refocus is for the purpose of enhancing successful accomplishment of Goals

IMAGERY

AN **INDIVIDUAL** WITH A **VISIONARY GOAL** THAT HAS THE ABILITY TO **MOTIVATE TEAM MEMBERS** TOWARDS ACCOMPLISHING **YOUR VISIONARY GOAL**

Thought is Impossible without Image ARISTOLE

Imagery: A mental skill to enhance goals Accomplishment

SCIENCE: KNOWLEDGE OF IMAGERY

Definition   Why   How   When

Definition Imagery: An imagined visual that imitates reality, which serves a purpose. Imagery is from the Latin *imago* meaning "imitation or copy"

The brain and the eye work together to process images by converting them into specialized electrochemical signals, then storing these components in the Occipital Cortex for future recall. (Kotulak, 1997)

Allowing the brain to recall pictures, create visuals, and add senses is a function of our brain's memories from past experiences.

Blind at Birth: There is one exception to the human brain storing pictures in the memory. A blind individual, born blind, has never witnessed a picture for the brain to store. Blind at birth individuals dream not in pictures but in the senses they possess. Their remaining senses are developed far beyond individuals with sight.

Imagery is not the same as daydreaming. Daydreaming is unintentional and does not serve a purpose. <u>Imagery is intentional and serves a purpose.</u>

OH, SO THAT'S WHAT YOU MEAN!

In the 1960s I became a student and a practitioner of mental skills. Several years later, my wife confessed she could not see pictures in her mind. My stumbling attempt to convince her fell on deaf ears.

Several years later, in a conversation we were having about our children when they were small, she made the following statement. "I can still see Stephanie (our daughter) trying to learn to walk in our back yard when she was still wearing a diaper and her pretty pink top she loved so much." At this point I allowed her to continue without interruption. "I can still see her falling down then getting back up; she just kept trying until she could finally walk without falling."

I replied, "What did you just say?" She repeated that she was able to see her, as if it were yesterday. My elated response, "That's imagery!" Her enlightened expression indicated she finally understood the meaning of *imagery* (Oh, so that's what you mean!). Being able to pull up visuals and senses from the memory is Recall Imagery. Aristotle summarized imagery most effectively, "*Thought is Impossible without Image.*" Not only does thought create image, each thought creates its own unique chemicals.

The right hemisphere of the human brain is our imagery center. Just as physical and technical skills can be improved with practice, the same is true with the mental practice of *imagery*. Once you develop the skill of *imagery* it becomes a life skill available upon demand.

- **Recreational Imagery** Two Purposes:
  Recalling special moments
  Creating special moments
- **Performance Imagery** Two Purposes:
  Recalling Performance
  Creating Future Performance

RECREATIONAL IMAGERY

Recall Imagery: Creative Imagery

Imagery involves visuals and may also include movements and other senses from the brain's memory. In the HOW section of *imagery*, a simple exercise is presented, that is dependent upon your memory recall.

Creative Recreational Imagery (advanced skill): Individuals with advanced imagery are able to create visuals and motion then store these created pictures and movements in the memory for future use.

Creative imagery may be thought of as pieces of a puzzle the brain has stored. Then it selectively chooses puzzle pieces to create a new image with movement and added senses.

Often times novice learners of *imagery* take it much too seriously resulting in negative imagery accompanied by a high level of stress-frustration. If necessary at first, just think about it and later you will see the images. "Thought is impossible without image." Have fun with practicing and developing *imagery*.

Example: Recreational Imagery: Combined Recall with Creative Imagery

THE PURPLE SWEATER

Imagery: Fun and Exciting!

At the age of 17, I attended Southern Illinois University to pursue my bachelor's degree. Sitting in my not-so-exciting Psychology class, we were participating in a memory recall exercise. Before our professor could ask my fellow colleague John the next exercise question, a female class member, arriving late, walked across the front of the class, and sat at the desk in front of John. She was very well blessed and wore a skin tight, bright purple sweater.

Undisturbed by this late arrival, our lady professor proceeded with the exercise quickly pointing to John and said, "NAME A COLOR!" Straight faced, without hesitation, and remaining seated, John's stature elevated a good four to six inches with a sudden uncontrolled contraction of his gluteus maximus. Excitedly, he answered with a loud shout, "P u r p l e." It took some time for the laughter to recede. Our professor could have called upon any of the male students with "name a color" and would have received the same exuberant testosterone response. Be careful of your recall imagery in public. Sitting here more than a half century later, my recall imagery can still pull up the purple sweater.

There is no doubt my brain has created (creative imagery) more to the purple sweater than actually existed when I was 17. Imagery can be and is a lot of fun, and in John's case, exciting. The purple sweater is not meant to be sexists or purple profiling; it is simply reality. It is very possible some younger male readers might think the Purple Sweater belongs in the category: *Why Imagery Is Important*.

PERFORMANCE IMAGERY

Recall Imagery and Creative Imagery

Recall Performance Imagery is no different than Recall Recreational Imagery with the exception; its purpose is to evaluate a past performance.

Creative Performance Imagery is no different than Creative Recreational Imagery with the exception; its purpose is to improve a future performance. Elite performers, in all professions, use creative imagery to improve self-confidence in an upcoming performance.

Rehearsing in front of a mirror is an excellent exercise in developing *imagery*. One would be hard pressed to find athletic officials who haven't rehearsed signal jesters in front of a mirror for future successful performance or business a man or woman rehearsing in the same manner for an important presentation. Any media in which one can observe visuals is an asset for practicing and developing *imagery*.

PERFORMANCE IMAGERY RESEARCH: (RECALL AND CREATIVE)

Basketball Free-Throw Shooting

Equally talented basketball free-throw shooters were randomly divided into three equal groups.

> Group A: Shot 100 free-throws daily for 45 days
> Group B: Shot 100 free-throws daily for 45 days using imagery only
> Group C: Didn't practice free-throws during the 45 day research

Results

> Group A: Improved 24% (practiced shooting)
> Group B: Improved 22% (imagery only)
> Group C: No improvement (no practice)

The results of this imagery experiment are an important reason WHY imagery is very beneficial for your leadership and your team member's performance goals.

WHY: IMAGERY

Imagery is practiced to assist in the success of team member's performance goal. The development of imagery is a systematic process, requiring focused self-discipline. The memory of the human brain is the storage of still and moving pictures plus our remaining senses.

An example: What is the color of your vehicle? _____ The process in which you have knowledge of its color is stored visuals in your brain. If you think about it you can see it. Should you choose to change the color of your vehicle in your mind's visual, you have exercised creative imagery.

Scientific Research: The Institute of Neurology in London, England, found at least 80% of the same brain waves are used during imagery as compared to the actual physical experience. With proper exercise it is believed this 80% may be greatly increased.

TWO SCIENTIFIC THEORIES

Psychoneuromuscular Theory: This scientific theory suggests imaging certain actions produces the same path for neuromuscular messages that are transported via the right hemisphere of the brain to the correct muscles. Psychoneuromuscular response provides an individual a more automatic response in the same or like situation.

Symbolic Learning Theory: This theory creates a "coding" of certain actions into a symbolic element, allowing a more automatic response in like circumstances. One might choose to think of repetitive imagery practice as developing a path in the brain's function for future application. Perceiving brain paths may be likened to building a highway infrastructure. You may find a better cogitative perception personal to you in your thoughts of how imagery works.

GET ALONG LITTLE DOGGIE GET ALONG

Each time thinking of developing a path in the brain, I involuntary recall the following experience as a young boy. Raised on a ranch/farm in Southern Illinois, it was my, my pony's and my dog's assigned duty to round up the cattle each evening and drive the herd into the barn corral for the night.

After a lengthy and repetitive practice of herding the cattle, a path in the pastures would start to develop and become visible. Thereafter, as I and my two four-legged comrades started the cattle drive most of the herd would automatically start towards the path and follow it to the open gate and into the corral. However, occasionally one or two of the cows would walk past the open gate.

Get along little doggie is analogous to developing brain paths and brain gate openings for the mental skill of *imagery*. There will be times when you use *imagery* the visual will not be exactly as you would have liked. In other words, you may have a couple of cattle walk past the open gate. *Imagery* is not an exact science, but we are getting closer.

APPLIED: PERFORMANCE IMAGERY (RECALL AND CREATIVE)

Possibly They Are Just Lousy Free Throw Shooters

Our basketball team was shooting 71% from the free throw line prior to the Christmas break. The won-loss record was dismal. As a team they averaged shooting between nine to twelve free throws per game. The reason this team shot few free throws per game is the players hesitated to penetrate the opponent's defense in fear of being fouled. The players' lack of self-confidence in going to the free throw line was obvious.

I approached the head coach in the following manner, "Would you mind if I worked with our kids on free throw shooting over the Christmas break?" He replied, "Go ahead, but it's possible they are just lousy free throw shooters."

With the proper practice of imagery, these kids could greatly improve their free throw shooting percentage. Working with the first five, each was filmed shooting free throws from four positions: front, both sides and back. (Filming is a tremendous asset in developing imagery)

Viewing the films, we identified technical errors and made technical corrections. Also proper techniques were identified for imagery practice.

Each player was given his personal film to view, on a daily basis, for two weeks. During that time the players practiced shooting free throws using imagery only for a minimum of 15 minutes an evening before sleep (prior to sleep the brain has entered the Alpha stage, Chapter 6 Relaxation). In using *imagery*, they focused on the corrected techniques and seeing themselves perform with self-confidence.

Result: The team's free throw percentage skyrocketed. The number of free throws shot per game skyrocketed. At the end of the season their percentage from the line averaged an amazing 92%. Their remaining schedule resulted with free throw performance statistics such as; 27 for 29, 33 for 34, 34 for 34, 35 for 36, etc.

The players no longer feared penetrating the opponent's defense. Self-confident players no longer fear being fouled, quite the opposite. With improved self-confidence in free throw shooting ability, they wanted to go to the line. The team's winning record reflected this improvement.

Advanced Imagery: These young basketball players, for the fun of it, practiced shooting some free throws with their eyes closed in daily practice. A game in which there was a substantial lead, one of the improved free throw shooters shot a free throw with his eyes closed and hit nothing but net.

The head coach turned and said with a look that could kill, "Please tell me I didn't see John shoot that XXX free throw with his eyes shut." Hesitating, I was searching for an out, but finally assured him he had seen it. The head coach looked away, tapped my leg and said, "Good job." Imagery improved technique and self-confidence in these team members. This is an example of Imagery empowering team members with self-confidence. Imagery assists in improving and accelerating the success of reaching performance goals within your team culture.

It is amazing the number of Division I collegiate basketball teams and professional teams eliminated from post season tournaments because of poor free throw shooting. The basketball free throw imagery results and various other imagery experiments is WHY to practice imagery

Numerous professions commit little to no practice time for mental skills. *In any profession the mental aspect is the difference between mediocrity and moving to a higher level of sustainable success.* In evaluation of your team culture is there a need for imagery in assisting successful goal accomplishment? In your own professional leadership role is there need for imagery? A major benefit of Imagery is the reduction of stress, when combined with the mental skill of Relaxation (Chapter 6). In your profession do you have a team or team members in key positions whose performance would be greatly improved with building self-confidence?

Why Practice Imagery: 3 major reasons

- To Overcome a Weakness in Performance, Injury, Illness
- To Improve a Strength
- For Enjoyment, Fun, and Stress Control

***The ultimate Why** in practicing and implementing the mental skill of imagery is to enhance the accomplishment of desired performance goals.

1. **To Overcome Weakness in Performance, an Injury, or an Illness**: In an earlier discussion, a weakness in performance (free throw shooting) was overcome in part by applying imagery

In Performance: As a doctoral student I witnessed the following transformation of performance in my university science class. Each student was required to give two presentations in front of the class. In my presentations I would always be in the top three of excellence. I was proud of this because in my earlier life I dreaded giving a public presentation of any kind.

One of our classmates gave his presentation and it was absolutely horrible. We were required to verbally critique each other's presentations. And so we did, offering ways for improvement.

Several weeks later it was this young man's turn to give his second presentation. We each were holding our breath in hopes he would improve his earlier performance. Improve! Improve! He knocked our socks off. We were all stunned. His performance was head and shoulders above all the rest of us.

My first thought was the only way this quantum leap in performance was possible was if he had brain surgery. Our professor called upon me to critique his performance. "That was a great presentation. How did you make such an improvement from your last presentation?"

He replied, "I took everyone's suggestions from my first presentation and practiced them in front of a mirror." Imagery is a most powerful mental skill. It can and at times does produce quantum leaps in performance.

Implementing imagery to eliminate weakness will definitely provide positive results. In Chapter 1 you made a self-evaluation of your character, *Who am I*. Can you recall a weakness you identified in your evaluation of *Who Am I*? On the scale of 0-9 how clearly can you image your weakness? (With 0-1 weak and 8-9 excellent, choose 2 consecutive numbers-0 1 2 3 4 5 6 7 8 9.) Think about seeing your weakness; then systematically image yourself reacting in a self-confident manner.

An Injury: Properly applying imagery accelerates healing. Mentally seeing the good cells in your blood overtake the bad cells will enhance and accelerate healing. Remember, at first if you can't see it, just think it. Eventually you will allow yourself to see the mental visual. This visual does not have to be anatomically correct to be effective.

Mental skills are interrelated. Relaxation with imagery to assist in the healing process will produce greater positive results. (Chapter 6 Relaxation) Applying imagery with physical rehabilitation is a formula for significantly accelerating and improving healing success. Mentally recall a personal physical injury such a turned ankle, a cut, a broken bone. Practicing imagery with relaxation could have accelerated your healing process.

Illness: Imagery to improve one's chances for success in overcoming an illness. Imagery is not a miracle. It is a mental tool to assist medical and spiritual treatment. Individuals who have developed imagery prior to their illness increased their chances for positive results more so as compared to those who just started development of imagery during illness.

2. **To Improve a Strength:** In Chapter 1 you made a self-evaluation of your character, *Who Am I*. Can you recall a strength you identified in your self-evaluation? If you wrote down your strength it was processed in your brain's memory. Regardless of whether it is a mental or physical strength, if it is not used it will deteriorate. Practicing your strengths is just as important and beneficial as practicing to overcome a weakness.

3. **For Enjoyment, Fun, and Stress Control:** Imagery is most effective when practiced with enjoyment, relaxation, and yes, fun. It is very possible your memory recall may produce an image from your past, rivaling or surpassing the Purple Sweater. Forcing imagery is counterproductive.

Evaluating Your Leadership:

- Can you think of a specific circumstance where you would use imagery? Example: Conference, Presentation, Team Meeting and etc.
- Now evaluate a team member within their positional boundary of your organizational team. Can you think of a specific circumstance where your team member would use imagery within their positional boundary?
- Think of a weakness for your team member to overcome: lack of self-confidence, inability to refocus, stress.

- Chose a team member's strength to enhance: A physical, technical or mental skill

Think of why you would use imagery in your personal life? Is there an area in your personal life you would sincerely like to improve? (Examples; weight loss, attitude, self-confidence, stress, composure, sleep, etc.)

Imagery is a powerful tool to use in pursuit of overcoming injury or illness (regardless the kind of injury or illness). Imagery assists in transforming a weakness into strength. The personal development of imagery becomes a life skill. It would be unusual to meet anyone who did not thoroughly enjoy and appreciate the science and art of imagery in their professional and personal life.

ART: APPLICATION OF IMAGERY

HOW: TO PRACTICE IMAGERY

Imagery practice results are best achieved by employing the KISS Method (Keep it Simple Stupid), learn to crawl- learn to walk- learn to run. This is one of the best methods of **HOW** to practice imagery

Simple Imagery Exercises: Thought is impossible without image

- Think of a circumstance which was an enjoyable and meaningful experience to you? Can you recall your experience?
- Think about your chosen experience. Can you see it? It's there; your memory has made available the stored pictures and possibly other senses.
- Write down a number between 1 and 10. Preferably make the number bold and larger than you normally would. _____Later you will be asked to recall this number from your brain's memory.

Imagery | 211

- Can you see your number in a different color? If so what color did you change your number?_____Your ability to create a different colored number is creative Imagery.

- What is the color of your vehicle? _____ Can you start your car and hear it running? (Yes No) Can you see yourself shutting your engine off? (Yes No)

- Think of a food extremely delicious to you? (Example Chocolate)_____ Using imagery place your chosen food in your mouth. Can you experience its scrumptious, delicious taste? Was this mental imagery experience realistic enough to physically excite your taste buds?

- In your mind's visual, think of a place special to you. This special place should be where you can be very relaxed and feel good about yourself. This is your very personal place of tranquility. It might be in a closet. It might be on a beach or possibly in the mountains. This environment is your personal choice. Can you see this place in your mind's visual?

Previously you were asked to record a number between 1-10. What is your number? Record your number here _____. Were you able to recall your number? If you were successful it is because you retrieved the image from your brain's memory. If you were unable to recall any of the exercise answers, it is simply because of your stress level. (Controlling your stress level Chapter 6 Relaxation)

*Common Misperception: Some novice practitioners of imagery may perceive the results they should achieve are to be identical to seeing a picture hanging on the wall or likened to a movie. In imagery you create and see what is important to you.

Presently, the human brain is not able to multi task, being able to complete two thought processes at exactly the same time. However, we certainly have the ability to quickly go back and forth from one task to another. While practicing imagery think of just one thing important to you. Think it, feel it, see it.

Important Tools in Your Development of Imagery: Mirrors and film are very powerful tools in developing and practicing imagery. Chose any profession and it would be difficult to find an individual whose performance has not been practiced in front of a mirror or viewed in a film.

MIRROR, MIRROR ON THE WALL

Who is the worst baseball hitter of them all?

As a young baseball pitcher in high school I was what one might call, hitting challenged or just a horrible hitter. In college it was necessary to develop my hitting techniques.

Solution: I bought a hitting instruction book, opened the text with pictures, and placed it on a stool in front of a mirror. I practiced the techniques, watching myself in the full length mirror numerous times prior to the next game.

Result: Three triples, driving in five runs in four times at bat. My hitting success continued. When I would go into a minor slump I would go back to the basics and practice my technical performance in front of a mirror. The same success holds true using films, which can be controlled by stopping, reversing and slow motion.

Remember, imagery is developed from pictures and movements stored in your brain's memory. This technique in developing imagery is very common in many personal performances in life. Being able to see yourself in a picture, film, or mirror is of great assistance in practicing and developing new neurological pathways in the brain for improving your imagery skill.

How: to practice imagery is important. The positive results of imagery are the greatest when you learn to believe in them. Mike Shannon, the great announcer for the St. Louis Baseball Cardinals, became overly excited during a close game and misspoke the following statement, "Folks, you have to believe it to see it." What he meant to say was, "Folks, you have to see it to believe it." When Mike Shannon misspoke he adequately described how to practice imagery, believe it and you will see it.

The skill of imagery requires your positive thoughts in creating belief. Your expectations will influence your positive imagery results. A large percentage of circumstances in leadership and in life will be determined by your expectations. With only a few exceptions, your expectations will determine outcome.

EXERCISE FOR YOU

EXPECTATIONS

Washer on a String

Several years ago in a media broadcast Deepak Chopra was interviewed. Dr. Chopra has studied and researched the power of the human brain. During his appearance he presented a washer (approx. size of a quarter or half dollar) attached at the end of a string (12+ inches long). With his elbow sitting on a stationary object he held the string between his thumb and forefinger with the washer about 1-3 inches off the table. With his mind he thought (imaged) the washer to do certain things. **First,** he stated he wanted the washer to start swinging back and forth, it did so. **Next,** he thought about seeing the washer circling clockwise, it did so. **Finally,** he imaged the washer on the string to reverse itself to move in a counter-clockwise movement, the washer performed as expected. Hesitatingly, the hostess finally conceded to try. Once again the washer performed as it had with Dr. Chopra.

AUBURN UNIVERSITY

Mental Skills Program: Football

Presenting a mental skills program to the football team at Auburn University I demonstrated the washer on a string exercise. With my elbow planted on the glass screen of an overhead projector, the football squad was able to witness the experiment. The success or failure of the experiment was visible on a large drop down screen.
Result: At first nothing … knowing full well the washer would start moving the way as stated, I made the following comment for suspense, "I have not practiced this in some time and I am feeling a little pressure here. I don't know if it will work."

Then it visibly started moving back and forth. Next, imagining the washer to circle and then reverse the circle, it did so. The exercise performed perfectly.

Knowing there would be some doubters: Not a problem. One of the doubters was chosen to come up front and perform the exercise; he accomplished the same positive result.

The washer on a string may be referred to as a poor man's lie detector machine. Each works on about the same principle. This exercise in thought and imagery really works. On the surface of our skin there is minute muscle activity called the Galvanic Skin Response (GSR). Invisible to the eye, it is there. It may be perceived as your thoughts and beliefs traveling across the surface of your skin. The washer on a string reacts to your GSR. Everyone has the mental skill of imagery and is capable of developing this skill to a higher level. (Martens, 1997)

Now it is Your Turn: You will need two things

A washer about the size of a quarter or half dollar or a ring
A string at least 12 inches long or more

Tie the object (washer or ring) to one end of the string. Hold the other end of the string between your thumb and your first finger. Place your elbow on a flat surface such as a desk or table. With your elbow planted extent your hand forward a few inches so the washer hangs free and above the flat surface. With your other hand steady the washer, and then remove your support hand.

Mentally think about the suspended washer starting to move back and forth. It may take a few seconds (5-20+), but it will start. While the washer is still moving, think about the washer starting to make a circular motion either counter or clockwise. Now think about the washer reversing its circular motion. At first, it may take several seconds, but the washer will slow and reverse the circular motion. If at anytime during the experiment you do not see the results you wish, then close your eyes and just think about what you want the washer to do.

*Your expectations determine outcome, such as the washer on a string exercise. Your expectations also play a major role in developing your mental skill of Imagery.

Twice previously you were asked to record a number between 1 and 10. What is your number? Record your number _____. You were able to record your number because you could see your number stored in your brain's memory. One more time; try to image your number with a little more clarity or vividness. Do this with your eyes open or closed. It is what works best for you. Should you so desire, have your number change color. Experiment with this exercise and permit yourself to enjoy and have fun with your incredible brain practicing *imagery*.

How to Practice and Apply Imagery

Presented are the Do and Don't of how to develop your imagery skill. Keep in mind the practice of *imagery* is systematic. You are creating pathways for the electrical impulses in your brain.

Do:

* **Do see Yourself Performing with Power**: Practice seeing yourself performing a (speech, technique, presentation, etc.) with power. You are focusing on your physical demeanor. Carry yourself with strength and enjoyment. I have yet to find anyone who loves being around a grumpy or negative individual.

* **Do see Yourself Performing with Enjoyment**: Practice seeing yourself perform with enjoyment. This will assist in seeing positives of your imaged performance.

* **Do see Yourself Performing with Self-Confidence**: A self-confident performer does not fear failure. Practice seeing yourself reacting positively to a negative occurrence. Self-confidence will allow your imaged performance to be more effective. Self-confidence is one of the major qualities present when a performer experiences flow.

* **Do see Yourself Performing in Slow Motion and in Reverse** (advanced): Once you have systematically practiced imagery you may now practice advanced imagery. Practice seeing yourself performing in slow motion. Play around with this skill. See yourself in reverse. Bo Schembeckler the famous University of Michigan football coach taught many skills of this game in reverse. Coach Schembeckler would have an offensive lineman demonstrate the finality of an assigned block of his opponent. Then he would have this team member move backwards one step at a time to the origin of the skill. In this case back to the players three point stance prior to the snap at the line of scrimmage.

Advanced: Do add senses to your *imagery*: If you are imaging a speech presentation and you have a favorite red tie, a favorite blue blazer, a favorite suit or a favorite dress, add these images to your practice. Place yourself in a performance position important to you. Design your image for what is important to you.

Question: When you dream do you dream in color or black and white?

ART: APPLIED CREATIVE IMAGERY

RELAY HANDOFFS IN MEN AND WOMEN'S TRACK AND FIELD

In practicing blind handoffs for track and field sprint relays, each relay team practiced the following: Our runner in the handoff zone would practice receiving the imaginary blue baton from the imaged teammate. Our runner would image hearing running footsteps and breathing from imaged runner approaching the handoff zone. Waiting for the baton at the assigned position in the exchange zone, the runner would image hearing his teammate yell "Go" at the designated time.

Once the baton was exchanged (our runner has the baton throughout this exercise) the skill would be completed by running through the exchange zone with competitive acceleration and speed. This imagery exercise was highly successful with both male and female track members. Men and women's 800 relay teams set school records.
*The principles of this *imagery* example are applicable with any individual seeking to improve performance.

Inside-Outside Imagery: Some may see *imagery* as being an observer watching themselves from the outside as a spectator. While others may see the imagery from the inside looking out. It is possible to practice and develop both inside and outside imagery. Favoritism of choice is decided by what works best for each individual.

USSR FILMED OLYMPIC VILLAGE INSIDE-OUTSIDE IMAGERY

Prior to the 1984 Olympics, held in Los Angles, the Soviet Union's mental skills coaches came to the United States and filmed the Olympic Villages where their athletes would be assigned. They also filmed the route from the Olympic Village to the appropriate venue where each of their teams would be performing. The mental skill coaches also added crowd noises at the appropriate times.

The Soviets applied this same procedure prior to the 1976 Summer Olympics. (Williams, 2009) In viewing the video of the Olympic Villages, the athletes would create images of themselves performing in their respective venues.

The USSR Olympic athletes would view the appropriate films applicable to their individual performance arena. Having viewed the film several times each individual team member would practice imaging themselves leaving their assigned room in the Olympic Village, traveling to their venue, and imagining their performance. Each team member was enabled to perceive they had been there before. Consequently, each team member was able to control his or her appropriate stress level. (Chapter 6 Relaxation) This mental strategy allowed leaders and team members to become more self-confident.

A more *confident* team member is a more *competent* team member. The objective of this *imagery* procedure was to build self-confidence. Soviet team members were permitted to practice either inside imagery or outside imagery performing their technical skills, with emphasis on precision. Which is best inside or outside imagery? Many performers develop both.

Question/Answer: When I practice or use *imagery*, should I do so with my eyes open or shut? Try both ways; it's what works best for you.

Don't:

* **Don't Practice Negative Imagery:**

Caution: Negative imagery is as powerful as positive imagery and produces negative results. (Williams, 2009) Your stress level is inadequate. When you first start your practice of *imagery* and you do not achieve desired results you have two choices in attempting to correct this issue:

1. Use a personal relaxation technique such as the 10-count (Chapter 6). If you were successful in controlling your stress level you will now experience a positive *imagery* practice.
2. If you are still not successful, walk away and continue at a later time.

*If you try to force this issue you will start in a downward spiral leading to an elevated stress level.

* **Don't take the Fun out of Imagery:** Design enjoyment and fun into your *imagery* practice. At this point many readers (both genders) most likely have created an enjoyable image rivaling or surpassing the Purple Sweater.
* **Don't see Yourself Finishing First:** It is best to see yourself achieving your performance goal.

Dr. Shane Murphy former head sport psychologists at the United States Olympic Training Center in Colorado Springs, Colorado, made the following statement: "When one practices imagery to see themselves finishing first, they are taking away from what they are trying to accomplish."

***Remember**, your performance is about you; not your competitor. You have no control over your counterpart; however you do have control over yourself.

* **Don't Force the Practice of Imagery:** Experienced practitioners, on occasion, will be dissatisfied with their practice session of *imagery*. The mental skill of relaxation is beneficial in this circumstance.

WHEN: TO PRACTICE IMAGERY

In practicing *imagery*, each individual will be at a different level of success. Some will be very successful almost immediately. Some will be successful with a few practices. Some will take longer to view success. Should an individual not be satisfied with degree of success, it is simply because that individual has yet to develop a proper mental technique for controlling stress level, (Chapter 6 Relaxation).

When novices struggle with *imagery*, it is primarily because they are not consciously recognizing the stored pictures in the brain's memory. The visual is present in the brain; it is a matter of permitting yourself to acknowledge the pictures. Pictures and the other senses are the totality of the brain's memory. It is impossible to have thought without images, with the exception blind at birth.

Providing the correct physical and mental environment is important in the beginning process of learning when to practice *imagery*. Your physical environment should include a private, personal, uninterrupted quiet surrounding. Prior to sleep the brain is in the *Alpha* stage. This is the perfect time as to when one may practice imagery. Your ability to develop a relaxed mental environment on demand will be thoroughly discussed in Chapter 6 Relaxation.

When: To Apply Imagery (Just about anytime)

WEST POINT CADET: APPLIED IMAGERY

While serving a Doctoral Internship at West Point I had the opportunity to pick the brain of many of the cadets who were involved in the Performance Enhancement Training Center. This performance training program focuses on the mental skills of goals, imagery, relaxation, and focus/refocus. The program includes a complete scientific laboratory of technological equipment including *Alpha* chambers. (Alpha chamber is an egg shaped, partially enclosed recliner designed for evaluation and practice of mental skills.)

The mental skills training program at West Point is one of the best in the world, as is the U.S Olympic Training Center. In a state-of-the-art program this technological equipment is a necessity for a successful program of this magnitude. The technology defines what is and what is not occurring in the development of mental skills. However, in our daily lives expensive technology is not necessary. Inexpensive technology is available to the individual.

Talking with the captain of West Point's baseball team, I discovered two mental aspects of this highly trained cadet. First, he had been offered a position on the United States Olympic team. (At this time Baseball was an experimental event) He chose not to participate. I asked him, "Why?" He looked me straight in the eye with an immediate response, "My goal and my focus is to be the best I can be in serving my nation in uniform. Being in the Olympics would interfere with my goal."

He was also the leading hitter on West Point's baseball team and his grades were perfect. He continued by explaining how he balanced playing baseball with his grades. "When I cannot take hitting practice because of my studies I spend at least 15 minutes after lights out using *imagery*. I see myself perform the proper techniques in hitting. I never see myself hitting a single or a home run. I see myself using the proper techniques in making contact with the ball." (Performance Goal)

Can you use transference from this West Point Cadet's example of knowing when to use imagery as being beneficial in your leadership and in your personal life? The science and art of human behavior is a wonderful and an exciting necessity for elevating your successful, sustainable leadership to its highest level.

Scientific research significantly suggests *imagery* is most effective when the mind is in a relaxed phase, the *Alpha* stage. We all experience

the *Alpha* brain waves prior to sleep. The West Point cadet applied his hitting *imagery* in the evening prior to sleep.
When: To use Imagery

APPLIED IMAGERY

Quarterback Fumble

Early in the football season our team was playing on the opponent's field. On first possession and the third play, the first string quarterback was knocked out of game with injury. Our second string quarterback entered the game never having taken a snap with the first team in a game situation.

On the first three offensive plays, our young quarterback fumbled handing the ball to the running backs. After taking the snap from the center he would raise the ball about a foot then lower it again (hitch) before delivering the ball to the belly of the running backs. This technique allowed perfect timing with the second team backs which were not as fast as the first team backs.

This technical flaw was explained to this young performer. He needed to eliminate the hitch in presenting the ball to the quicker first team backs. While his defense was on the playing field, this young athlete took a few steps behind his teammates and practiced his *imagery* skill in delivering the ball to the backs with the proper technique.

Leaders know it is difficult to break a habit quickly in a performance situation. His skill of *imagery* worked. No more fumbled hand offs the remainder of the contest. In this example imagery was implemented to correct a weakness in a technical skill.

Prior to the season and during the season this young athlete had practiced his skill of *imagery*. It is not practical to think the skill of *imagery* or any of the mental skills as an event; it is a process.
*The application of *imagery* may be implemented almost anywhere, anytime.

When: Not to Use Imagery

During Technical Execution: Using the previous example, it would have been counterproductive had this quarterback exercised his *imagery* while taking the snap from under center. His focus was to take the snap, not on using *imagery* to deliver the ball to the running backs.

* Developing the mental skill of *Imagery* is to enhance the success of performance goals.

AN **INDIVIDUAL** WITH A **VISIONARY GOAL** THAT HAS THE ABILITY TO **MOTIVATE TEAM MEMBERS** TOWARDS ACCOMPLISHING
YOUR VISIONARY GOAL

Chapter 5 Imagery

SUMMARY

Accomplishing Your Visionary Goal
with Team Member Imagery

Imagery is a powerful mental skill. Understanding and applying the combination of the mental skills **imagery** and relaxation may produce results, which in some cases presently are unexplainable. Developing the mental skill of **Imagery** is to enhance the success of the ultimate, your visionary goal. In most instances **Imagery** may be more effectively practiced with enjoyment and fun.

The mental skills of Imagery, Relaxation, Refocus assists in the successful accomplishment of Your Team Member Goals

CHAPTER 5 IMAGERY: MAJOR TOPIC POINTS

Science: Knowledge of Imagery

What is Imagery: A mental visual imitating reality

Two classifications of Imagery
- Recall Imagery: Visualizing a past event
- Creative Imagery (advanced skill) Creating a new visual

Two purposes of Imagery
- Recreational
- Performance

Recreational
- Recalling special moments
- Creating special moments

Performance
- Recalling Performance
- Creating Future Performance

Why Imagery is Important
- The Purple Sweater

Two Scientific Theories: (pathways of the brain)
- Psychoneuromuscular Theory
- Symbolic Learning Theory

Art: Application of Imagery

How to Practice Imagery: (visual learning)
- Mirror Film

Do:

    Power/Strength

    Enjoyment

    Self-Confidence

    Slow Motion and Reverse

Don't:

    Negative

    No Fun

    Finish First

    Force Practice

When to Practice Imagery

    Appropriate Physical Environment

    Appropriate Mental Environment

    Applied Imagery: West Point Cadet

When not to use Imagery

    During Technical Execution

CHAPTER 5 GLOSSARY: IMAGERY

Creative Imagery: Individuals with advanced imagery are able to create visuals and to create motion (usually a performance) then store these created pictures and movements in the memory for future use

Imagery: An imagined visual imitating reality which serves a purpose

Psychoneuromuscular Theory: This scientific theory suggest imaging certain actions produces the same path for neuromuscular messages that are transported via the right hemisphere of the brain to the correct muscles. This psychoneuromuscular response provides an individual a more automatic response in the same or like situation

Recall Imagery: An individual recalling visuals, which may also include movements and senses from the brain's memory

Symbolic Learning Theory: Creates a "coding" of certain actions into a symbolic element allowing a more automatic response in like circumstances. One might choose to think of repetitive practicing imagery as developing a path in the brains function for future application. You may wish to choose perceiving these brain paths as building a highway infrastructure. Or you may find a better cogitative perception personal to you in your thoughts of how imagery functions in your brain.

Mental Skills of *Imagery*, Relaxation, Refocus are for
the purpose of enhancing successful Accomplishment of Goals

Chapter 6:
Relaxation

◆ ◆ ◆

INTRODUCTION

AN **INDIVIDUAL** WITH A **VISIONARY GOAL** THAT HAS THE ABILITY TO **MOTIVATE TEAM MEMBERS** TOWARDS ACCOMPLISHING **YOUR VISIONARY GOAL**

Accomplishing Your Visionary Goal with Team Member Relaxation

*Goals   Imagery   **Relaxation**   Refocus*
The art of stress control is important in profession and has numerous benefits in personal life including life crisis.

Yoga, Zen and Hypnosis are not included in this discussion although each is also highly effective. The proper *relaxation* procedure will vary with each individual. What best works for you within your circumstance is the proper choice.

Progressive Relaxation and the *10-Count* are presented since each of these techniques may be used without an instructor and each may be used anywhere, anytime. *Relaxation* skill is a mental technique for successfully controlling one's stress level. *Relaxation* techniques play a vital role in successful accomplishment of professional performance goals and personal life performance goals.

The purpose for controlling stress is to achieve the proper level of stress, most often the *Alpha stage*. Scientists have learned those who experience highly successful performance have a sudden burst of *Alpha brain waves* just prior to the accomplishment. This important mental technique, *Relaxation*, may be developed by anyone with appropriate practice.

CHAPTER 6 RELAXATION: MAJOR TOPICS

SCIENCE: KNOWLEDGE OF RELAXATION

ART: APPLICATION OF RELAXATION

Guided Dreams

Nova 2009 Science: What Are Dreams

Chapter 6 Glossary: Alpha Brain Waves, Fight or Flight, Galvanic Skin Response (GSR), NREM, Relaxation, REM, Parasympathetic Nervous System, Sympathetic Nervous System

> Mental Skills of Imagery, *Relaxation*, Refocus are for the purpose of enhancing successful Accomplishment of Goals

Relaxation

ALPHA BRAIN WAVES

AN **INDIVIDUAL** WITH A **VISIONARY GOAL** THAT HAS THE ABILITY TO **MOTIVATE TEAM MEMBERS** TOWARDS ACCOMPLISHING **YOUR VISIONARY GOAL**

Relaxation skill is not a luxury; it is a necessity.
TRIATHLETE

Relaxation: A Mental Skill to Enhance Goals Accomplishment

"One can read the environment much more clearly when you are 'Calm Internally', just as you can see the reflection more clearly on a calm lake, than on a disturbed one. Stress is like the wind that disturbs the image on a calm lake." (Orlick, 1990, p.149) Meditation (relaxation) has its tradition grounded in Eastern cultures, but has been popularized the last several decades in Western cultures.

SCIENCE: KNOWLEDGE OF RELAXATION

Definition   Why   How   When

Definition Relaxation: Cognitive control of stress level

Relaxation is a mental exercise affecting the body processes. This allows an individual to mentally gain control of their psychological stress level. The mental skill of *relaxation* is not something new. To the contrary, the origin of *relaxation* practice is difficult to determine. However, it is quite evident Eastern Countries have understood the importance of *relaxation* procedures (martial arts) for centuries. Many of the mental skills we are presently discovering are most likely commonplace in countries such as India and Tibet. In fact, to this very date the Western World is discovering benefits of *relaxation* skills the Eastern World has known as far back as can be researched.

*Important: A certain amount of stress is needed. Without stress our desire to accomplish would be severely compromised. Without stress our *fight-or-flight* response would never be activated.

During a *relaxation* procedure a number of physiological changes take place:

Blood pressure drops
Heart rate drops
Breathing rate slows
Brainwaves of the cortex slow down thus approaching or entering the *Alpha stage*

It's all about Alpha: *Alpha* brain waves are one of four basic brain waves prior to and during sleep (*Alpha*, Delta, Theta, Beta). All four of these brain waves are oscillating electrical charges in the brain. Each brain wave serves a useful purpose.

Alpha brain waves are present in wakefulness where there is a relaxed and effortless alertness. *Alpha* waves oscillate 8-13 times a second.

Scientific theories suggest a flickering flame, sound of a waterfall, and a water fountain are relaxing because they possibly imitate the oscillating waves of *Alpha*.

Alpha brain wave benefits:

Enables less stress
Enables sleep
Enables improved strength of immune system
Enables peak performance in achievement
Enables peak performance in personal life

Regardless the kind of peak performance, there is a sudden burst of *Alpha* brain waves usually on the left side of the brain. (Hardt, 2005) Achieving *Alpha* brain waves is possible with different *relaxation* techniques. Our focus will be the application of *Progressive* and *10-Count relaxation* techniques to achieve the *Alpha* brain waves.

WHAT IS THE MENTAL SKILL OF RELAXATION

Relaxation techniques are a deliberate mental process for controlling one's stress level. *Relaxation* is not a kind of sleep, but a deeply focused yet resting state of mind. (Hayward, 1998)
Two Functions of the Central Nervous System for Controlling *Relaxation* (stress) Level

1. **Sympathetic Nervous System Function:** (Increasing Energy) Elevation of energy in the human body: The Sympathetic Nervous System function is distributing and <u>increasing energy (stress); causing the elevation of blood pressure, increased breathing, increased heart rate, and perspiration.</u>

2. **Parasympathetic Nervous System Function:** (Conserving Energy)

The parasympathetic system assists in returning the human body to a relaxed state after experiencing increased energy. The Parasympathetic Nervous System function is <u>conserving energy</u> (stress) via *relaxation* skills; causing the lowering of vital signs such as respiratory rate, heart rate, perspiration, etc.

Excessive stress (a high level of energy) may be a product of good things happening. (marriage proposal-that may be questionable, accepted new employment, pay raise, beautiful Purple Sweater, etc.) Also excessive stress occurs when the brain perceives a threat. The central nervous system then reacts by setting off a chain of chemical reactions. Elevation of energy is a function of the Sympathetic Nervous System. Our brain perceives a threat when it has been thrust into a situation it believes will have major consequences. This cognitive perception will trigger the *fight-or-flight response.*

Fight or Flight defined: Your body prepares itself, when confronted by a threat, to either stand ground and fight or run away.

Greensburg describes what happens when the human brain encounters a high level of stress. The *Sympathetic Nervous System*, activated by the hypothalamus, regulates the body to do the following:

Increase heart rate
Increase force with which heart contracts
Dilate coronary arteries
Constrict abdominal arteries
Dilate pupils
Dilate bronchial tubes
Increase strength of skeletal muscles
Release glucose from liver
Increase mental activity
Dilate skin and muscle arterioles
Significantly increase basal metabolic rate

Because of these physiological changes, the human being has been able to perform incredible feats in emergencies. The power of the *fight-or-flight* response is not yet fully understood. Once the *sympathetic nervous system* has completed its response, it then becomes the function of the *parasympathetic nervous system* to return the relaxed state as it was prior to being highly stressed.

Important Good News: Through biofeedback recent studies indicate the involuntary functions of the body are not totally involuntary. As the human brain evolves, it becomes possible to train cognitive functions to be available upon demand. Exercising and developing the mental skill of *relaxation* may be much more powerful than previously thought.

WHY RELAXATION

Hunting the Elusive Zone-Alpha: *Relaxation* techniques are crucial to your professional and personal health. Without some form of *relaxation* technique our mind, professionally and personally, may become like the winds of a developing tornado. Learning to control stress through *relaxation* techniques is likened to the calming of the wind.

Reasons for the practice and development of your *relaxation* skills are numerous. In leadership, in team members, and in our personal life *relaxation* skills produce important benefits which include personal, mental, and physical health. Earlier research thought stress of major life changes was the most harmful; however recent research indicates everyday stress is more harmful to health than those occurring in major life changes.

Everyday hassles are more detrimental to health than major life changes. When stress is chronic, prolonged, or goes unchecked the <u>results are illness and disease.</u> Research significantly indicates uncontrolled stress slows the healing process. Developing a mental technique for stress control is important in all aspects of leadership and personal life.

It is important you develop successful, sustainable leadership without destroying the rest of your life. *Relaxation* is crucial in this achievement.

Leaders and team members who pursue goals have a distinct motivational advantage in successful accomplishment of the mental skill *relaxation*. Goals provide purpose and incentive to discover new techniques in going beyond past performance. More so now than any other time in many Western Cultures, leaders and team members are discovering the benefits of imagery, *relaxation*, and refocus in pursuit of excellence.

Learning *relaxation* techniques is a big part of learning to "Keep Cool" under pressure. (Murphy, 2004) Relaxation research has demonstrated we are in greater control than previously thought in our ability to control stress. Control is brought about by developing the mental skill of Relaxation combined with Imagery. Combining these two important mental techniques will be discussed in the How section (Relaxation Skills).

Mental Gas Pedal Stress is necessary. Under differing circumstances *relaxation* skills are designed to control the appropriate level of stress. Your ability to regulate your body's stress level with mental skills of *relaxation* is analogous to your ability to control your vehicle's speed with the gas pedal in different speed zones in traffic situations.

Combining *Imagery and Relaxation*

One technique many have found to be useful is to combine *Imagery with Relaxation*. A mental technique for *relaxation* may be thought of as the deliberate development of a mental gas pedal to control the appropriate degree of stress. Mentally choose a vehicle, a sports car, an SUV, a Truck, any make, model and color of your choice. Now install your mental gas pedal into your imaged vehicle to regulate the speed of your stress.

Example: Apply imagery to see your chosen mental vehicle as a means of traveling the three stages of stress. You may control your stress level in accordance with your circumstance. On the following continuum of stress, traveling in your imaged vehicle, you may pick and choose which of the three stages of stress you desire.

Continuum of Stress for Performance: 3 Stages

1. (Not enough stress)__ 2. (**Alpha**-correct amount)__ 3. (Excessive stress)
Too Slow   **Maximum Performance**   Too Fast

* Ninety-nine percent of the time when we encounter stress it occurs in stage 3 or analogous to a vehicle going 90 mph in a 65 mph zone.

* Less than 1% of the time we encounter stress level to low for maximum performance. This is stage 1 on the continuum of stress or analogous to traveling 15 mph in a 65mph zone.

* Developing *relaxation* skills enables achievement of **Stage 2 Alpha** (maximum performance) on the Continuum of Stress. Traveling 65 in a 65 zone is analogues to *Alpha*.

___ 65 mph____ _____ 65 mph _____ 65 mph_____

(15 in a 65 zone)   (perfect 65/**Alpha**)   (100 in a 65 zone)

Example: Imagine you are in your mental vehicle in your driveway and about to drive to your destination in which the major number of miles is posted with a 65 mph speed zone. In driving to the 65 mph zone you will travel through several different speed zones. Entering this 65 mph zone you personally believe your driving skills will perform best between 60 and 70 mph. Returning home, once again you travel through posted speeds which require altering your speed using your gas pedal.

Developing a *relaxation* technique to control your stress level, for varying personal and professional circumstances, is analogous to having a gas pedal for controlling the speed of your vehicle in different speed zones and emergency situations. In developing your mental gas pedal you are controlling your sympathetic and parasympathetic nervous system.

WHY RELAXATION: (4) IMPORTANT BENEFITS

- *Achieve Maximum Performance* in Leadership and Life
- *Recover from Injury or Illness*
- *Refocus in performance* when things do not go as planned
- *Achieve the Brain's Alpha Stage* (Alpha brain waves)

*Sustainable Maximum Performance in Leadership and Life: requires mental skill of *Relaxation*

1. Achieve Maximum Performance: Often we think of sport when we mention maximum performance. However, this is a false analogy. There are important performances in each profession; such as a presentation, a speech, an acting performance (such as television, movies, politics, etc.)

Think of a special time in your profession in which you believe maximum performance would be very beneficial? Most of us are visual learners. Recording your thought is important for recall imagery.

Remembering the *Who Am I*

Important performances are often present in our personal lives. Maybe it's a special evening you have planned to be with someone. Perhaps it's a special day or event for your family or a special moment in time. Ensuring success in our private lives involves the pursuit of maximum performance while staying true to our values. Think of a special time in your personal life in which you believe maximum performance would be very beneficial. Recording your thought is important for recall imagery.

UNIVERSITY OF TEXAS: FAMOUS FOOTBALL COACH (1956-1976)

Achieving Maximum Performance with the Mental Skill of Relaxation

A good friend asked for assistance in implementing a mental skills program at a private educational institution in Austin, Texas. After being on campus for a few days, one of the school's administrators made a point to pick my brain concerning the mental skills program.

As I started to elaborate on the mental skill of *relaxation*, his demeanor (kinesis) changed as he seemed delighted to relate his past experience with the mental skill of *relaxation*. This large-statured male administrator (former offensive lineman at UT) began proudly relating how his Division I football team (early 1970's) was encouraged to exercise the mental skill of *relaxation*, prior to each game.

Coach Darrell Royal believed the practice of *relaxation* was beneficial prior to performance in pursuit of performance excellence. Modern technology has proven Coach Royal to be correct. His players' brain waves entered the *Alpha Stage* prior to their performance on the football field enhancing *Maximum Performance*.

Prior to performance, 99% of the time, when one's stress level is not in Stage 2 (Alpha), it is in Stage 3 (too much stress). Excessive stress prior to performance is comparable in attempting to fill an ice cube tray with the faucet on full volume. The water splatters everywhere and results in failing to fill the tray. The *Alpha Stage* is analogous to controlling the faucet water volume to successfully fill the ice cube tray. Achieving Stage 2 on the stress continuum allows maximum blood and oxygen flow to muscles, the brain and to different functions of the body enabling *Maximum Performance*.

Coach Royal's leadership provided the skill of *relaxation* to enhance his team members' ability to enter the *Alpha Stage* in pursuit of *Maximum*

Performances. This is true of any performance. Coach Royal's won/ lost record validated his belief.

The leadership of Coach Darrell Royal provided the following results during his tenure at the University of Texas (1956-1976): Three National Championships and Eleven Southwest Conference titles. This legendary leader was plainspoken, honest and had a sense of honor. Success with Character: Good guys do finish first and often.

2. Recover from Injury or Illness: Scientific research has made it <u>impossible to deny</u> the positive interaction between mind and body and the effects of one upon the other. It is a definite, excessive stress inhibits patient's recovery. Research also indicates the mind may increase the body's susceptibility to disease. This connection is referred to as *Somatogenic.*

Individuals who overreact to stress from injury, disease, or treatment handicap their chances of a speedy recovery. Furthermore, excessive stress produces *cortisol*, which triggers hunger, causing weight gain. Ongoing scientific research studies the body's chemicals for communicating between the mind (nervous system) and the body (immune system). Developing the mental skill of *relaxation* is extremely beneficial in combating injury, disease, and enhancing medical treatments success. (Greensburg, 2010)

For example, pretend you have just sprained your ankle. Your ankle starts to swell due to blood rushing into the area around the injury from ruptured blood vessels. Depending upon the severity of the sprain, your trainer or physician most likely suggests you apply cold several times a day for duration of 24 to 48 hours. Applying cold helps to slow the rushing blood and consequently controls the degree of swelling.

Next, it is appropriate to apply heat. Applying heat will break up the dried blood in the swelling. Now over time your circulatory system will eliminate the dried blood causing the swelling to disappear. This process may be accelerated with relaxation skills. Individuals who are experienced in the art of mental skills may combine *relaxation with imagery* for best results.

Relaxation skills improve recovery time. Improper psychological stress hinders the effectiveness of our immune system. Excessive stress will hinder recovery because it constricts or handicaps the proper flow of your circulatory healing process. (Stage 3 on the stress continuum) *Relaxation combined with Imagery* allows your circulatory healing process to perform near or at its maximum level. (Stage 2 Alpha on the stress continuum)

U.S. NATIONAL FOOTBALL LEAGUE RUNNING BACK

Implementing a Mental Skills Program at a Division I university, I had the honor of working with a young student/athlete, one with excellent character. He would always sit in the middle of the second row during the mental skills presentations. Without exception he was always very attentive. It was his goal to play in the NFL upon graduation. By his kinesis, I could tell he was absorbing every mental concept presented.

During the presentations of *relaxation and imagery* I made the following comment, "Those of you who will someday play in the NFL will most certainly be challenged by injury. No one escapes injury in professional football." This young athlete was in the top five of the NFL draft. Sure enough, he played running back in the NFL and was chosen rookie of the year. Early into his second season, he sustained a career ending injury, so everyone thought including his coach. In the healing process he utilized his mental skills.

Eventually, he recovered and today is a running back in the NFL. After his return to action his coach made the following comment, "I do not know of anyone else who could have returned from such an injury." His mental attitude, his trainers, his doctors, his deep belief in his Creator, and his mental skills are credited with return to his loved profession. There is no doubt to this day we do not fully understand the incredible benefits of the mental skills *Relaxation* and *Imagery* when combined.

Illness Recovery: Medical Treatment

"Developing the mental skill of *Relaxation*, combined with specific *Imagery*, results in the production of substances, Neuropeptides." (Greensburg, 1990, p.40) They are beneficial in fighting disease and enhancing success of medical treatment.

Patient recovery research indicates a positive significant difference when patients believe they have some control over the treatment and healing process. This belief of control enables a *relaxation* level (*Alpha*) for accelerated healing. Numerous studies link guided *Imagery* combined with *Relaxation* techniques with lowering stress during medical treatments, including surgery, chemotherapy, and MRI.

Imaging healing does not have to be anatomically correct. Mentally see white blood cells in your blood vessel ingest and destroy bacteria and virus. Even though this may not be anatomically correct it is effective in developing belief. Patients, who believe they have some control over healing, accelerate the healing process. *Relaxation* combined with *Imagery* increases the development of belief.

3. Refocus in Performance: (When Things Do Not go as Planned) A complete discussion on recovery in performance will be presented in the mental skill of refocusing. It is beneficial to remember in any performance, whether leadership, team member, or personal life; *Inability to refocus is the number 1 cause of failure* (Chapter 7 Refocus).

ART: APPLICATION OF RELAXATION

4. Achieve the Brain's Alpha Stage: How to Practice *Relaxation* The proper *relaxation* procedure will vary with individuals. What best works for you is the proper choice. The following discussion presents two *relaxation techniques;* Progressive and 10-Count. In personal experience of helping others develop the mental skill of relaxation, it is most productive to first learn *Progressive Relaxation*.

Progressive Relaxation Script (modified)

First, get yourself into a comfortable position (Lying down, comfortable chair)

Think and see into your right arm and hand

Squeeze your right fist and hold it for the count of three

Gently open your hand and focus on the relaxation in your right hand and arm. (Mentally repeat Relax, Relax, Relax)

With each repetition of relax–mentally say relax more slowly

If you so desire repeat this exercise one or two times more

With your mind's visual think yourself into your left arm and hand

Squeeze your left fist for the count of three

Gently open your hand and focus on the relaxation in your left hand and arm (Mentally repeat Relax, Relax, Relax)

With each repetition of relax, mentally say relax more slowly

You may continue in any order you choose until you have relaxed the entire body, if you haven't fallen asleep. In the future you may also wish to include focusing on your breathing during your *relaxation* exercise.

*Important: Do not get up quickly after progressive *relaxation exercise.* The blood in your body is not evenly distributed. You should sit up slowly and then stand up slowly. Some describe the experience of the uneven blood distribution as a sinking sensation while others may describe it as a floating sensation.

The manner in which the former football player at the University of Texas described their pregame relaxation procedure was a modification of *Progressive Relaxation* just described. The practice of *Progressive Relaxation* is a good foundation for any other relaxation techniques you may wish to practice.

10-Count

May be implemented anytime anywhere. It is important to reiterate this procedure is most effective after you have practiced *Progressive Relaxation* for several days.

The following is a script for the 10-Count procedure:

- Mentally see and say (eyes closed or open)
  10 Relax, Relax, Relax

  9 Relax, Relax, Relax

  8 Relax, Relax, Relax

- With each mental repetition of relax (slow down) your thought process
- You may also wish to focus on your breathing
- Stop this procedure when you have reached your desired level of relaxation

After practicing and understanding you are able to control your stress level, it is beneficial for you to experiment with and develop your own personal procedure. Many great performers have done so.

National Basketball Association: Elite Star

An elite NBA basketball athlete developed a personal procedure when at the free throw line, especially when free throws were crucial. He would see himself beaming into outer space looking down on earth and realizing this attempted free throw was just a small piece of the puzzle of planet earth. This worked for him in controlling his stress level, a mental gas pedal for controlling stress. The 10-count method also allows controlling stress within a few seconds.

Your personal desire to discover the benefits of *relaxation* is important. Suppose you go to your favorite restaurant and order your desired food on the menu. Let's say you choose to order a steak dinner. Your order is brought with a beautiful presentation. It is sizzling and the aroma more than satisfies the desires of your taste buds. Just as you start to take your first bite an unexpected event occurs requiring you to leave, never having the opportunity to experience this wonderful steak. This analogy holds true with one having the opportunity to practice and develop the many benefits of *relaxation*, then deciding for whatever reason not to do so. Relaxation Skills are no longer thought of as a luxury for those who wish to pursue and elevate excellence in leadership and life.

WHEN RELAXATION

Just about anytime, anywhere

A Simple Exercise Right Now: (sitting in a chair) Here is a simple exercise you may execute now to achieve a physical relaxation sensation. If you are sitting in a chair place your left hand to hold the left side of the seat bottom of your chair. Do the same with your right hand on bottom right of your chair. Now try to lift the chair you are sitting on with your hands and arms only. Pull hard for the count of 1001, 1002, 1003; now slowly relax your hands and arms. Did you feel the *relaxation* in your hands and arms? If you did not feel this sensation you simply need to pull harder with your arms. Should you so desire repeat this exercise focusing on your *relaxation* sensation.

*Elite performers in all professions have been found to exhibit a sudden burst of *Alpha* brain waves prior to successful performance. These elite performers being able to control their stress level is a product of dedicated self-discipline.

Guided Dreams

Three actual experiences of Guided Dreams occurred while practicing Relaxation combined with Imagery.

Implementing a Mental Skills Program with an Illinois Track and Field team the student/athletes religiously practiced setting goals, *imagery, relaxation*, and refocusing skills. Approximately two thirds through the track season, unexpectedly, unusual experiences occurred on several occasions involving different student/athletes. Each of these young athletes experienced guided dreams the night before their competition while practicing *relaxation and imagery*.

GUIDED DREAM #1

(Dawn) But I have never finished first!

Dawn a senior had participated three years in the 100 meter dash. On this occasion prior to the 100 meters Dawn approached and said, "I had the weirdest dream last night." When I asked her what it was, she responded, "I could see the runners on each side of me." She identified which teams were beside her and also the lane in which she would be running. She continued, "I dreamed I finished first in the 100."

Being my first year to work with this team, I asked Dawn, "Well what is weird about that?" She reminded me she was a senior and had never finished first.

Immediately the last call for the 100 meter runners was announced over the speaker system. As Dawn departed for the starting line I commented, "Well let's go see what happens."

Dawn was assigned lane 3 as she had seen in her dream. The gun fired for the start. As they came out of the starting blocks all the runners were in a dead heat. I had positioned myself close to the finish line. As they approached the finish Dawn and the runners in lanes 2 and 4 were neck and neck ahead of the rest of the runners. Approximately five meters prior to the finish line, Dawn gained a half meter advantage. She finished in first place exactly as she had seen in her dream the night before. Just prior to sleep each evening (*Brain's Alpha Stage*) Dawn listened to her *relaxation* script I had recorded for her and the team.

Dawn's father, an FBI agent, had been gathering inside information for the successful prosecution of a drug cartel. She informed her father of her successful guided dream. Because of his recent highly stressful assignments he started playing Dawn's *relaxation* program throughout their home's media system prior to bed time.

GUIDED DREAM #2

(Wayland) Time 1:59

Wayland, a second year 800 meter runner, had a guided dream after exercising his relaxation program prior to sleep. He dreamed about his race for the next day. The team was aware of the weather forecast, which was the possibility of light showers. Wayland's best time in the 800 was 2:08. He described his dream as follows: "On the second lap of my 800 meter race it started to rain and I finished my race with a time of 1:59."

The following day while Wayland was running the second lap of his 800 meters it started to rain. He finished third. The finish line timer who was responsible for third place walked up to Wayland to inform him of his third place finish and his official race time. As the timer approached, Wayland said, "1:59?"

The timer looked at his stop watch, "1:59." Being somewhat surprised the timer asked, "How did you know your time?"

Starting to walk away wearing a confident smile Wayland replied, "I dreamed my race last night and my time was 1:59."

GUIDED DREAM #3

(Psycho) Something Wasn't Right

Psycho (a name his team members affectionately assigned him) a senior was undefeated in the mile. Psycho would run his final race of the season. Prior to his mile, Psycho approached, needing to tell of his guided dream the night before. He said, "I saw myself performing very well, but something wasn't right. I couldn't clearly see myself on the last 200 meters." I asked Psycho what he meant: something wasn't right. "I don't know, but something wasn't right."

On the fourth and final lap of the mile on the home stretch Psycho was beginning to understand, "something wasn't right." His best friend and teammate, who had never beaten Psycho, opened up a 7 meter lead coming down the home stretch. A couple of meters before they crossed the finish Psycho barely edged out a first place finish to complete his career undefeated in the mile.

To this day I believe Psycho's best friend eased up just enough for Psycho to have an undefeated season. "Something wasn't right, now I know what that was," Psycho lamented.

Excited, I may have discovered something new in guided dreams. I sought information at the university library. **My finding:** Reports of guided dreams have occurred for many years. It appears this mental dream most commonly occurs when performers are involved in mental skills including performance goals, *imagery and relaxation.*

Nova 2009 Science: What Are Dreams

European, Canadian, Australian, and American university professors have extensively studied human sleep. During sleep our brain cycles in and out of NREM (non rapid eye movement) and REM (rapid eye movement.)

These prestigious researchers believe during NREM sleep, dreams are pieces of the past in which an individual attempts to assemble the dream. REM sleep dreams are in the future. They are creative. Since guided dreams of performance are creating the future it is logical to think guided dreams occur during REM.

Phil Jackson Levitate

Individuals who do not understand mental skills often have fun at the expense of those who do understand and practice these skills. Such is the case of Coach Phil Jackson's players. Phil Jackson is well known for his belief in *relaxation* procedures prior to competing and also during certain performance situations. Opposing players during pregame warm-up would often ask Coach Jackson players if they have already completed their levitation exercises. This is all in good fun and appropriate.

A REFORMED DOUBTER

In a basketball pregame locker room talk, I gave my players our strategy and their positional assignments and expectations. Then each player started developing their written performances goals in their goals journals. At this point our team record was 18 and 2. We finished at 22-2.

Unknown to myself, the head football coach whose staff I had been with two months earlier, walked into the locker room and stood where he could not be seen. As soon as the last player left the locker room for the pregame warm-up, the head football coach stepped out and said, "Hey, why didn't you do that with my kids?" (Referring to mental skills training)

My response, "I did ask you coach; your reply was, 'not on my time'." I immediately rejoined my team. You too will experience reformed doubters.

Each of my players prior to pregame was encouraged to listen (with head phones) to music they believed was relaxing to them. This was their personal choice in attempting to achieve *Alpha* prior to performance.

Today more and more leaders and performers are acknowledging and accepting the importance of mental skills. Relaxation is valuable in completing the big picture of mental skills training. In most instances your *relaxation* procedure is attempting to achieve the *Alpha Stage*.
*Relaxation techniques are all about achieving the *Alpha State* in accomplishing desired results in professional performance and personal life performance.

Affordable Technology

Relaxation recorded scripts: Numerous relaxation recorded scripts are available on the internet, in book stores, in retail stores, and other such sources.

Modern technological devices: Equipment for monitoring your ability to effectively use your mental gas pedal, for *relaxation*, is available at a reasonable price. The GSR2 is such a device, which may be purchased from *Thought Technology Ltd. Montreal, Canada*. The GSR2 resembles a computer mouse and is reasonably priced. Other sources may be located on the internet. Technology for monitoring your ability to manipulate skin temperature and other human functions are also available.

* Developing the mental skill of *Relaxation* is to enhance the success of *Performance Goals*, which ultimately achieves success of *Accomplishing Your Visionary Goal*.

AN **INDIVIDUAL** WITH A **VISIONARY GOAL** THAT HAS THE ABILITY TO **MOTIVATE TEAM MEMBERS** TOWARDS ACCOMPLISHING
YOUR VISIONARY GOAL

CHAPTER 6 RELAXATION

SUMMARY

<u>Accomplishing Your Visionary Goal</u> *with Relaxation Techniques*

Taking time to be alone with your thoughts is not a luxury, it is a necessity
Developing the Mental Skill of *Relaxation* is one the most important things an individual can do for themselves in professional and in personal life. Developing a mental technique to achieve the **Alpha** brain waves has numerous benefits which will vary, in importance, with each individual.

CHAPTER 6 RELAXATION: MAJOR TOPIC POINTS

Science of Relaxation: Knowledge

What is Relaxation: Controlling one's stress level

Alpha Brain Waves
Sympathetic Nervous System function

    Function to increase energy
Parasympathetic Nervous System function

    Function returning to relaxed state
Why Relaxation is Important

    Achieve maximum performance in leadership and life
    Accelerate recovery of injury and disease
    Assist in Refocusing
    Achieve Alpha brain waves

Art of Relaxation: Application

How to Practice Relaxation: Two Types

    Progressive with provided script
    10-count with provided script
When Relaxation should be Implemented

    Just about anytime, anywhere
Guided Dreams

    Nova 2009 Science: What Are Dreams
    Coach Phil Jackson and Levitation
Affordable Technology: GSR2 + others; where to purchase

CHAPTER 6 GLOSSARY: RELAXATION

Alpha Brain Waves: Present in wakefulness where there is a relaxed and effortless alertness. Alpha waves oscillate 8-13 times a second.

Fight or Flight: "Your body prepares itself, when confronted by a threat, to either stand ground and fight or run away".

Galvanic Skin Response (GSR): Skin's electrical field causing minute muscular activity which may be controlled by ones thought process.

NREM (Non Rapid Eye Movement): One of the brains cycles during sleep. Dreams during NREM are believed to be pieces from the past.

Relaxation: Cognitive control of stress level

REM (Rapid Eye Movement): One of the brains cycles during sleep. Dreams during REM are believed to be pieces creating the future.

Parasympathetic Nervous System: Assists in returning the human body to a relaxed state after experiencing increased energy.

Sympathetic Nervous System: Function is distributing and increasing energy (stress); causing the elevation of blood pressure, increased breathing, heart rate, perspiration

Mental Skills of Imagery, Relaxation, Refocus are for the purpose of enhancing successful Accomplishment of Goals

CHAPTER 7:

REFOCUS

◆ ◆ ◆

INTRODUCTION

AN **INDIVIDUAL** WITH A **VISIONARY GOAL** THAT HAS THE ABILITY TO **MOTIVATE TEAM MEMBERS** TOWARDS ACCOMPLISHING **YOUR VISIONARY GOAL**

Accomplishing Your Visionary Goal with Team Member Refocus

GoalsImageryRelaxation**Refocus**

The number one cause of Failure in Leadership and Life: Inability to Refocus

A broken focus in leadership and life performance is inevitable. How to *refocus* is the most difficult of all the mental skills. Thus, it becomes the number one cause of failure for desired outcome. Elite performers in all professions have developed a deep-seeded belief in which they perceive a broken focus as an opportunity for a learning process, enhancing success towards accomplishing the future. Regardless how one *refocuses* it always involves a *mental trigger*. In the How section of this chapter, the art of mental triggers will be discussed in detail.

CHAPTER 7 REFOCUS: MAJOR TOPICS

BRIEF DISCUSSION OF FOCUS

Science: Knowledge of Refocus

Why Refocus
When to Refocus

Art: Application of Refocus

How to Refocus
Mental Triggers

Chapter 7 Glossary: *Boxmentalize*, Devolution, Focus, Mental Trigger, Refocus

Mental Skills of Imagery, Relaxation, Refocus are for the purpose of enhancing successful Accomplishment of Goals

REFOCUS

MENTAL TRIGGERS

AN **INDIVIDUAL** WITH A **VISIONARY GOAL** THAT HAS THE ABILITY TO **MOTIVATE TEAM MEMBERS** TOWARDS ACCOMPLISHING **YOUR VISIONARY GOAL**

Goals   Imagery   Relaxation   **Refocus**

When you develop the skill of refocusing your knowledge becomes invaluable, not only to yourself, but also many lives you touch

Refocus: *A Mental Skill to Enhance Goals Accomplishment*

FIRST A BRIEF DISCUSSION OF FOCUS

Focus Defined: The ability to attend to proper stimuli during performance (Cox, 2011)

Focused individuals make far fewer daily decisions than unfocused individuals. Thus focusing conserves mental energy. Attempting to focus on more than one stimulus limits ones performance.

Presently, scientists are discovering new ways to tell when the brain is properly focused for certain performances. Broadly speaking, the brain generates four kinds of patterns: Delta (seen most often during sleep), Theta (when you're daydreaming or catnapping), *Alpha* (often observed when you are aware but relaxed), and finally Beta waves (the key one for cognitive processing). (Cloud, 2011) The latest technology called "Body Wave" can determine when your focus or refocus has peaked; thus knowing when you are primed for performance decision making.

The concept of focus includes the ability of a performer to both narrow and broaden focus attention. (Cox, 2011) According to Williams (2009), focus is always moving. For example, in football the quarterback has a narrow focus for proper hand placement under the center. As soon as the quarterback starts the snap count, the proper focus widens to the responsibility of reading the opponent's defense. When the ball is snapped a narrow focus is necessary to secure the ball. Immediately after the snap a broad focus is required to hit the proper receiver. On a running play a fumble on the handoff requires the quarterback to immediately achieve a narrow focus to secure the fumble.

Each thought developed in the quarterback's brain produces proper chemicals for the rapid response of opening and shutting appropriate brain gates. Our every thought produces its own unique chemicals for proper performance functions. (Kotulak, 1997)

Dr. Amir Raz McGill, Chair of Cognitive Neuroscience Laboratory University of Montreal, Quebec, Canada, has conducted numerous studies concerning focus. His research strongly indicates our brain in its present state is capable of focusing on only one stimulus at a time. Presently, multitasking is not possible if one considers the definition of multitasking as processing two or more stimuli at exactly the same time. What does occur, the brain switches back and forth between stimuli; such as the case of the quarterbacks' performance requirements just mentioned.

It stands to reason when a focused individual makes fewer decisions, this reserves brain energy by not switching back and forth between unnecessary stimuli. Some quarterbacks are highly successful in reading the different and changing defenses more so than some of their counterparts. Is it possible highly successful performers, which require quickly changing focus from narrow to broad, do so at a faster rate than those who are not as successful?

A former military officer, presently an educator, was presented the following question, "What is the number one quality brought to the classroom from your military training?" The reply was, "Being able to think on my feet, making quick decisions with differing circumstances." This response describes ability to focus and *refocus*.

Devolution: Brain's development regressing

In the future if the brain evolves, multitasking may become a possibility. Some theorists think the brain is in a devolution stage. These theorists point out earlier civilizations have accomplished enormous feats we cannot duplicate with modern technology.

Examples: The Great Pyramids of Giza, Egypt; Hanging Gardens of Babylon, Iraq; Temple of Artemis, Ephesus; Statue of Zeus at Olympia; Mausoleum of Halicarnassus; Colossus of Rhodes; Light House at Alexandria and other such ancient accomplishments

However, regardless the direction of our brain's development, it is inevitable one's focus will be interrupted by other stimuli. Once this occurs the problem becomes getting back on track with the desired stimuli. Getting back on track (*refocusing*) is a very difficult process.

The ability to narrow and broaden focus is necessary in numerous performances in both profession and personal life. The ability to focus quickly is a product of setting performance goals. Written performance goals are one of the best methods in developing the ability to *refocus*. When one sets a challenging, realistic, and written performance goal, the goal itself becomes the focus and often the *cognitive trigger* for *refocus*. Cognitive strategy to *refocus* is presented in the HOW section of this chapter.

SCIENCE: KNOWLEDGE OF REFOCUS

REFOCUS

Definition Why When How

Refocus Definition: The ability to regain maximum performance after a negative or positive distraction. Refocusing requires understanding and accepting what you can and cannot control.

Performance distractions (unwanted stimuli) appear when we least expect them. They come in two different categories, positive and negative. How an individual mentally processes distractions from their goals is the difference between success and failure. Of all the mental skills, *refocusing* is the most difficult and requires the greatest self-discipline.

Refocusing enables you, your position leaders, and team members to thrive in times of ambiguity and uncertainty, where those who fail to *refocus* find failure. (Sullivan & Harper, 1996) Elevating your leadership to its highest level requires a self-disciplined attempt to "make every obstacle an opportunity." (Armstrong, 2001)

FOCUS/REFOCUS EXERCISE

Focus: Regardless of the numbers involved, pair everyone. One person is the leader and the other the follower. Facing each other, both individuals place their hands about chest high with leader and follower palms facing each other about one inch apart. The leader then very slowly starts moving one or both hands. The purpose is for the follower to focus on the leader's palms and mimic the leader's movements. At some point have the leader and follower switch roles.

Refocus: Should you so desire during this exercise, walk between the two partners. Some may respond verbally, some may ignore you in successfully keeping their focus. You may want to be selective who you choose in attempt to break focus. While implementing this exercise during a

Psychology class, at the University of Southern Mississippi, I made the mistake of walking between two giant offensive linemen in attempt to break their focus. One of the offensive linemen picked me up by both of my scrawny arms and set me aside without saying a word, thus keeping his focus. Other techniques used in attempting to break focus are words, pictures, movements and sounds.

Suggested movie for Refocus: *Top Gun*

Caution: This movie may be offensive to some audiences since it includes swear words. A section of this movie is very relevant for observing *refocusing* and the failure to *refocus*. In *Top Gun* a dog fight occurs between American pilots and the enemy. A couple of American pilots, in the heat of battle performance, maintained their focus. One of the pilots *refocused* and one failed to *refocus*. This dog fight situation gives clarity and meaning to the importance of *refocusing*. *Top Gun* or other similar movies demonstrate *refocusing*. A movie clip such as this is beneficial for *Implementing Your Goals Program* explained in Chapter 4.

WHY

DEVELOP YOUR ABILITY TO REFOCUS

Number 1 cause of Failure in Leadership and in Life:
Inability to Refocus

Positive thoughts produce positive chemicals, while negative thoughts produce negative chemicals in the body's functions. When an individual cries, the tears of sadness have different chemistry than tears of happiness. An individual's ability to stay positive is important for our body's chemistry when performing. Where the mind goes; the body will follow. Often times when focus is broken an individual's stress level is accelerated. Relaxation is an important technique in controlling stress level and consequently ability to *refocus*. The ability to *refocus* enables an individual or a team to resume maximum performance.

Sustaining Successful Leadership

If you have several years experience in leadership it is very possible you know of an average or less than average leader who was successful for a limited time in spite of their leadership skills. However, you most likely cannot think of a leader with average leadership ability who sustained successful leadership. One of the major purposes of developing *refocus* is to sustain your successful leadership at its highest level.

> *The hardest part is not becoming a champion; the hardest part is staying a champion.*
> WAITE HOYT
> *New York Yankee Pitcher*

Waite Hoyt, a famous New York Yankee pitcher (1921-1930) compiled a sustainable winning record of 276 wins with 70 losses. After the Yankees had won another World Series a reporter asked Waite, "How hard was it to become a champion?" Waite Hoyt replied, "The hardest part is not becoming a champion, the hardest part is staying a champion." And so it is in leadership and in life, achieving sustainability. In Waite's situation a positive circumstance (winning the World Series) threatened to break focus (satisfaction) for continued success. How did he and his team *refocus* for sustained success? They focused on performance (intrinsic motivation) verses winning (extrinsic motivation).

Why: Refocusing Ability Sustains Success

The product of *refocusing* is achievement of sustainable, successful performance. Refocusing is not an inherent human behavioral trait. Developing the ability to *refocus* involves a conscious effort. Every great leader and every great performer have practiced and utilized *refocusing* on numerous occasions. When successful, the result is sustainability.

Great leadership understands that each team member's behavioral reaction to good and bad performance is exactly how they will react to good and bad circumstances in their personal life. Only a major intervention in life, such as death or a spiritual experience etc., will prove the preceding statement incorrect. If you were to look back in time when you were in your teens, would you see any differences in yourself between then and now, in how you respond to critical issues?

There are simply no words to effectively express the immeasurable importance of teaching each team member the ability to *refocus*. Before you can accomplish this you must first demonstrate your ability to *refocus*.

Should your leadership display inability to *refocus*, so too will your team members. However, just because you display *refocus* ability is no guarantee your team members will. When you the leader display *refocus* ability it creates an environment for your team members to do the same.

EXAMPLES: REFOCUS AND FAILURE TO REFOCUS

Refocus Applied: What do I do, Doc?

While serving on a Division I Athletic program the following awkward and unfortunate situation occurred. In a hubristic attempt to bypass NCAA rules, the President and Athletic Director of the University, in the dark of night, sought to hire a head coach to replace the present coach during the season. Their attempt to keep this illegality from going public failed. Consequently, the present head coach called a family meeting discussing this situation in which they may have to suddenly relocate.

The following day the head coach called me into his office and with a solemn demeanor asked the following question, "What do I do, Doc; how do I handle this situation?" In essence he was asking how to *refocus*.

I presented the coach the following advice, "Sincerely figure out in your mind, over what it is you have complete control and over what you do not have control. Focus all your mental energy on what you can control and do not spend one ounce of energy on what you cannot control. If you do waste energy on what you cannot control you will be cheating your players, your coaching staff, your family and yourself." He followed this advice to the letter and successfully *refocused*; consequently his tenure outlasted the University President and Athletic Director.

In this situation the coach *refocused* by developing a mental dialog (mental trigger) of placing all his mental energy on what he could control. Following this *refocusing* advice wasn't easy. Refocusing seldom is. This is an example of *Why developing the ability to refocus* is important in achieving sustainable success in profession and life.

Even the most hardened practitioners of *refocusing* will, on a limited basis, experience failure to *refocus*. The science of our own personal human behavior is an ongoing and ever changing process.

TWO PORTRAITS OF REFOCUSING

A POLITICIAN

"The sense of obligation to continue is present in all of us. A duty to strive is the duty of us all. I felt a call to that duty. The path was worn and slippery. My foot slipped from under me, knocking the other out of the way, but I recovered and said to myself, it's a slip and not a fall."

1831 Lost his job

1832 Defeated in run for State Legislature

1833 Failed in business

1834 Elected to State Legislature (Success)

1835 Sweetheart dies

1836 Had nervous breakdown

1838 Defeated in run for State House Speaker

1843 Defeated in run for nomination to U.S. Congress

1846 Elected to Congress (Success)

1848 Lost re-nomination

1849 Rejected for land officer position

1854 Defeated in run for U.S. Senate

1856 Defeated in run for Vice President Nomination

What about You? At this point with two successes and eleven failures, would you have found a way to refocus and continue on?

1858 Again defeated in run for U. S. Senate
1860 Lost Presidential debate
1860 Elected President of the United States (Success)
.....................Abe Lincoln

Why was it important for Abe Lincoln to *refocus*? In many people's opinion, he was the greatest public figure in American history, a rare and compassionate humanitarian. Abraham Lincoln suffered a multitude of failures and setbacks in his life, which he successfully overcame. He continually *refocused* and consequently won the presidency, changing the course of his nation's history. Abe was asked how he was able to come back from failure (*refocus*). President Lincoln used a mental perception (slip not a fall) as a *cognitive trigger for refocusing*.

Temporary defeat hardly needs to mean ultimate failure. Abe Lincoln developed the ability to *refocus*, to continue on regardless of circumstances. Developing the ability to *refocus* in leadership and in life requires a mental commitment and self-discipline. In Leadership and in Life getting knocked down is going to happen. What is important is how you get up (*refocus*).

AN OLYMPIAN

Ultimate *Refocusing*: Olympian Greg Louganis

Considered the greatest diver of all time, Greg Louganis won gold medals at the 1984 and 1988 Olympic Games on both the springboard and platform. He is the only male and second diver in Olympic history to sweep the diving events in consecutive Olympic Games. This feat is even more impressive when considering he was the favorite to win both events in 1980, but due to President Jimmy Carter's boycott of the Olympics, he was unable to compete. (Refocus?)

Louganis won an unprecedented 47 national titles during his career, in addition to three NCAA titles. From 1982 to 1988, he won five world championships, six Pan American Games gold medals, and four Olympic gold medals. He was the first diver to receive perfect 10s on a dive in international competition when he did so at the 1982 World Championships.

At the 1988 Olympics one of the most impressive athletic feats in sports history occurred on live television during the springboard preliminaries. Hitting his head on the diving board, Greg fell lifelessly into the water. Those at pool side and those of us watching on live T.V. wondered if this mishap would be crippling or fatal. Slowly, Greg emerged to the surface and gently made his way to the edge of the pool. Holding the back of his head, Greg Louganis and his lifelong coach Sammy Lee departed to the dressing room. His wound required several stitches, all of this coming from an athlete who prepared each performance with mental skills and technical precision.

Shortly thereafter, as Greg was leaving the swimming complex, the concerned news media were gathered for a hopeful interview. The first reporter asked him how he was. "I'm o.k.," Greg replied.

Then the reporter asked the following question, "What went wrong?"

Listen carefully to Greg's response: "Just a minor error in judgment, that's all, just a minor error in judgment." It was obvious by this comment, he was already in the process of *refocusing*, or was he? Greg Louganis's next dive was near perfection; winning the Gold medal.

Using the psychological skill of transference, how would you react if you were to hit your head on the diving board in the performance of life? Many of us already have. If so how did you *refocus* or did you? If you had a second chance, how would you respond in turning a failure into a great accomplishment? How would you attempt to *refocus*?

Leadership's Gold Medals

Watching Greg's final dive and winning the gold medal, I found I was unable to verbally express this wonderful refocusing experience. Remembering what a good friend once said, "The most beautiful things in life cannot be seen or touched; they are felt by the heart." Have you had the privilege working with team members with whom you shared successes and failures, shared tears of sadness and tears of joy, shared great performances, and saw the fruits of *refocusing*? Is it possible each one of your team members is your gold medal? The ability to *refocus* is a necessity for successful, sustainable performance in leadership and life. When you provide leadership with good character you will have numerous gold medals when you cross the finish line.

Portrait: Failure to Refocus

St. Louis Cardinals vs Kansas City Royals 1985 World Series

*Important Disclosure: To be fair to all the participants in this portrait, it is important to note in 1985, the term Refocus was not commonplace. Nor were the techniques of *refocusing* well defined or well known.

The Portrait: It was the second Missouri-only World Series: The first was the 1944 World Series between two St. Louis teams, the St. Louis Cardinals vs St. Louis Browns (a team which later moved and is now the Baltimore Orioles.)

The Cardinals were in pursuit to win their National League-leading, tenth World Series title, while the Royals were seeking to become the first American League expansion team to win the World Series.

The first two games were at Royals Stadium; the Cardinals won game one, 3-1, and game two, 4-2. The next three games were played at Bush Stadium. The Royals won game three: Royals 6, Cardinals 1. The Cardinals took game four, leading the Royals 3 to 0. At this point the Cardinals led the Series three games to one. All the Cardinals had to do at this point was to win one more game for the World Championship. However, game five belonged to the Royals 6 to 1. Still with a Series lead 3-2 the Cardinals had to win but one more game.

Game 6 started a sequence of errors; eventually causing the wheels to come off the Cardinal's wagon. At this point every call was of epic importance. However, soon to come would be a missed call in which the Cardinals lost their composure and their focus. *They failed to refocus.*

In the ninth inning of game 6 the St. Louis Cardinals had a 1-0 lead over the Kansas City Royals. The Royals led off the bottom of the ninth with a ground ball to Cardinal first baseman. He flipped the ball to Cardinal pitcher Todd Worrell, covering first, and clearly beat the runner. At this point the Cardinals were just two outs from winning the World Series. But hold on, the first base umpire Don Denkinger, an honest man and a good umpire, made a bad call ruling the Royals runner safe. Everyone watching the game in the stadium and on television knew the runner failed to beat the Cardinal throw to first base.

Instead of the Cardinals being two outs away from winning the World Series they now were three outs away. As the inning progressed Kansas City had runners on second and third with one out. The Cardinals chose to walk the next batter, loading the bases. Pinch hitter Dane Lorg lifted a single to right field, driving in two Kansas City runs and giving his team a 2-1 victory. The end of game 6 found the Series tied 3-3.

The Kansas City Royals celebrated this dramatic turnaround and mobbed home plate. The St.Louis Cardinals went to their dressing rooms in stunned disbelief, only to find champagne waiting for them and plastic covering their lockers in anticipation for the celebration. Umpire Denkinger stated he still believed he had made the right decision, until he later met with Commissioner Peter Ueberroth after the game and had the opportunity to see the replay himself.

The Stunning Coincidence: Umpire Denkinger was scheduled to be the home plate umpire in Game 7. Could the Cardinals put this unfortunate call behind them and *refocus*?

Kansas City homered off John Tudor in the second inning. The talented Cardinal pitcher Tudor left the game trailing 5-0 in the third. On his way through the dugout he hit a power fan, which resulted in a cut finger.

In the lengthy fifth inning the St. Louis Cardinals came completely unglued. A succession of five Cardinal pitchers allowed six Royals runs. Eventually, Cardinal Manager Whitey Herzog brought in volatile picture Joaquin Andujar, normally a starter. Andujar gave up an RBI single to increase the Royals lead 10-0 before things became truly nasty.

The St. Louis pitcher twice charged home plate umpire Denkinger to disagree with his strike zone. First, Denkinger called an Andujar pitch a ball, even though replays seemed to indicate it was a strike. Whitey Herzog emerged from the dugout to defend Andujar and was ejected, reportedly after saying to Umpire Denkinger, "*We wouldn't even be here if you hadn't missed the xxx call last night!*"

The next pitch was called a ball; the replay indicated a strike and Denkinger ejected Andujar after misreading a gesture to Tom Nieto. Andujar again lost his cool and charged at Denkinger. It took three teammates to restrain him and get him off the field. Andujar was suspended for the first ten games of the 1986 season for the outburst.

Kansas City would take the World Series with a 10-0 victory in game 7. The Royals became the first team ever to win the World Series after dropping Games 1 and 2 at home. The Royals also were the Sixth and (to date) the last team to comeback from a three games to one deficit to win the World Series.

Whitey Herzog later wrote he wished he had asked Commissioner Peter Ueberroth, who was in attendance, to overrule the call umpire Denkinger made at first base in the sixth game. If Uberroth had refused to do so, the St. Louis manager would have pulled his team from the field and forfeited the game.

Presented were a politician and an Olympian who successfully refocused. Also presented was a major league baseball team's failure to refocus despite having one of the greatest managers in baseball history. *The number 1 cause of failure in profession and life: inability to refocus*. Refocusing is required under extreme and important circumstances. Refocusing is difficult and requires composure, as well as great self-discipline in your leadership role. When you demonstrate the <u>inability to refocus</u> this creates a vacuum in your team culture. The preceding three portraits were presented to give clarity to: *Why refocusing* is important.

WHEN

AFTER A POSITIVE OR NEGATIVE CIRCUMSTANCE DISTRACTS YOUR FOCUS

In your leadership role you will be called upon to assist your team and your individual team members to *refocus* in different and difficult circumstances.

When refocus is necessary: More often than not *WHEN* to *refocus* is dictated by unexpected circumstance occurring in the following 3 situations

- Before performance
- During performance
- After performance

1. Before Performance

When team members set performance goals prior to a performance, it is seldom this team member will need assistance in *refocusing*. However, it does and can happen to any performer including veteran performers.

SPRINTER

Loss of Focus

Such was the occasion when our best sprinter prior to his 200 meter dash made the following comment, "Coach I have lost my focus; can you help me *refocus*?" I must admit I was caught off guard when this highly talented, undefeated sprinter lost his focus. Searching for a *mental trigger* I asked the following question, "What did you write down in your mental skills journal for your goal in the 200 meters?"

He replied, "I know coach, but that doesn't help."

I knew this young man's background so I asked, "Has your father ever seen you run the 200 meters?" He said he hadn't. "Would you like it if he could be here to see you run? Would it help you to *refocus* if you would use your imagery to see your father at the finish line of your race?"

"Yes coach that is what I will do."

Immediately after he embraced this *mental trigger*, the final call for the 200 meter dash. When the sprinters rounded the curve and entered the straightway four runners were neck and neck. With 40 meters remaining our runner accelerated and won by a large margin. "Thanks coach, it really worked for me." This highly talented athlete's mental trigger to *refocus*, prior to performance, was using imagery to create a meaningful cognitive perception.

2. During Performance (positive or negative distraction)

When: Do I *Refocus*? After a positive or negative circumstance breaks your original focus. More often than not when to *refocus* is determined by a negative circumstance during performance. However, positive distractions also occur during performance.

800 RELAY: SETS SCHOOL RECORD

Positive Distraction

At an invitational track and field meet our 800 meter relay team unexpectedly broke a long standing school record in the event. The 800 team approached jumping up and down in great elation. Before their celebration had ended, the last call for the low hurdles was announced. The situation was this: The young man who just anchored the 800 relay was the state's second fastest runner in the hurdler event.

Also in this event was the state's fastest hurdler. When our runner heard the last call for his favorite event his demeanor immediately changed to one of fright. "What do I do, coach; what do I do? How can I *refocus*?"

I looked him straight in the eye and said, "What did you write down in your journal?"(Mental Skills Journal) Can you mentally see the goal you wrote down for the hurdles?"

He started back pedaling towards the starting line while saying, "Got it, I've got it coach; that works." His *mental trigger* to *refocus* was to visualize his *written performance goal.*

The gun fired to start the race. As the runners came off the turn and entered the straightaway our runner and the state's fastest runner were in a dead heat. They remained that way through the finish line. From my vantage point it looked like a tie. However, the judges on the finish line called our runner for first place. This young performer was able to *refocus* from a positive distraction by mentally seeing his written goal. His *mental trigger to refocus* was imaging his written goal.

3. After Performance: Need for *Refocus*

One can find examples of positive distractions in every profession. For example in sport, a basketball shooter scores a three pointer and continues to admire the shot while the opposing team is at the other end of the court scoring. In football, a player or players celebrating before the game is over.

ARMY VS NAVY

Positive/Negative Distraction: After Performance

Several years ago the Army/Navy football game was played on national television. Navy was leading by two points with five seconds left on the clock and Army was moving the ball up field. Two seconds remained on the clock; Army was going to attempt a winning 38 yard field goal. The Army field goal kicker, an avid practitioner of mental skills, had earlier in the season kicked a 40 yard field goal. However, this was an Army/ Navy game with two seconds left. If the field goal was good Army won, if not Navy won. Pressure! The snap, the hold, the kick, and the ball split the uprights as time was about to expire. Great Jubilation!

But wait; there was a flag on the play against Army; off sides. Now the football was placed on the 43 yard line, three yards longer than his longest career field goal. The crowd was going crazy. Pandemonium!

Let's hit pause a minute: Now you are the Army field goal kicker. What is going through your mind? Can you *refocus* after thinking you kicked the winning field goal? Instead you realize your team has been flagged and now you are asked to do something you have never done before in your career. What would you be thinking?

Now back to reality: The snap, the hold, the kick ... the ball was up and on its way right through the uprights for an Army victory. How did this young cadet *refocus* under such tremendous pressure? I don't know the answer. But I do know he focused on his performance and not on the outcome.

Remember the words of wisdom from Jules, "Those who fail start thinking about the prize and forget about what it takes to get to the prize." This cadet successfully *refocused* under great pressure. I wonder what his mental trigger was for his successful field goal. *What mental trigger would you use* with a similar circumstance in your profession or in your personal life?

ART: APPLICATION OF REFOCUS

HOW

MENTAL TRIGGERS

Foremost, you must evaluate what you can control and then place all your energy on your control in Profession and in Life.

In the words of West Point's Colonel Csoka, "Know what it is you want and don't let anything get in your way of this accomplishment, while staying true to your moral character." This statement gives one direction and purpose in developing the ability to *refocus*.
Techniques for Refocusing: Mental Triggers

Mental Trigger Defined: Internal dialog to direct thoughts in a deliberate, positive direction

The big question is how do I go about *refocusing* in extremely difficult circumstances? Our brain works very efficiently with mental triggers. For example, when you think of a special food you love, such as chocolate ice cream or a delicious steak, does it trigger your taste buds? Most likely it does. The mental trigger to set off your taste buds is a thought, a visual of your favorite food. (Thought is impossible without image.) Developing and practicing your personal *mental trigger for refocusing* is very effective in your brain's cognitive process.

Triggers of the Mind: (In search of Alpha Brain Waves)

* A Goal (think and see your Written GOAL)
* A Word (such as mentally or verbally repeating focus, focus, focus)
* An Image (seeing with the mind's eye a beautiful lady or a handsome man)
* A Thought (about a loved one or a role model–something personal)
* A Touch (right hand to the left side of your chest-Beam me up, Scotty)

*Develop your very own unique technique

Ridiculous Triggers: Or are they ridiculous?

* Sacred space
* Phone booth
* Zoom to outer space
* Invisible shield
* Fly above the Clouds

*If it is ridiculous and it works, it is not ridiculous.

DEVELOPING YOUR PERSONAL MENTAL TRIGGER FOR REFOCUSING

In developing your very own unique mental trigger, remember the effectiveness and the value of ridiculous association. This mental technique is an excellent memory tool for *refocusing*. Our three pound grey matter remembers things better when they are unusual, vivid, very enlarged, humorous, sexual, or just plain ridiculous. In my experience of helping others, I have found the two best *mental triggers* for *refocusing* are ridiculous association and written goals.

In your leadership role think of an important situation in which your focus possibly could be broken. If you were asked to think of a ridiculous association *trigger for refocusing*, what would it be? Your thought: _____

When you develop your *personal trigger for refocus* it becomes more meaningful, effective, and exciting. Give your *refocusing* technique serious thought. Failure to do so should not be acceptable.

Great performers in all professions and life have three like qualities:

* Developed great self-confidence (performance goals)
* Do not fear failure (makes failure a learning process)
* *Developed the unique ability to Refocus (develop a mental trigger)*

Veteran performers eagerly perceive a negative situation as a learning process. Furthermore, they *Boxmentalize*, mentally place the negative circumstance in a box until after the performance or performances have been completed. Eventually, they open this mental box and study how to increase their chances of this negative not recurring. Their negative circumstance becomes a learning process for future successful performances.

Achieving belief is a deliberate mental process requiring self-discipline. Self-discipline is invaluable for developing *refocusing*. It becomes a personal choice to develop or not to develop the ability to *refocus*. Each of us has the cognitive ability to choose.

WHAT ABOUT YOU?

In your leadership role can you recall a circumstance in which you failed to *refocus*? In your leadership role can you recall a circumstance in which you successfully *refocused*? In your personal life can you recall a circumstance in which you failed to *refocus*? In your personal life can you recall a circumstance in which you successfully *refocused*?

If you answered yes to the preceding questions what do you think was the difference between failure and success?

98 PER CENTER OR 2 PER CENTER

Which One are You

As a young boy I remember my parents calling a conference at the dining room table. Their concern was I took life too seriously. I dearly loved my parents and I was grateful for their concern. Several years later I came to realize I wasn't taking life too seriously, instead I had a deep passion for wanting to understand human behavior. Presently, this deep passion is still very active.

In my teens, 20s, 30s, 40s, and even in my 50s I believed there were two types of individuals in personal study of human behavior. I sincerely believed in observation of *refocusing* ... 98% of those who were pursuing success came within an eye lash of a great accomplishment, then hit one more hurdle and said, "That's it; I can go no further." As a result they failed. Two per cent found a way to *refocus* and were successful in pursuit of their great accomplishment.

In the past 15-20 years those percentages have improved because individuals pursuing success are practicing mental skills. If I had to put a percentage on the *refocusing* success of today's individuals, instead of 98 percent failure, about 90% of those who pursue successful accomplishment *fail to refocus*. Instead of a 2 percent success rate today a 10 percent success rate in *refocusing* is experienced. This increased success is a product of mental skills and understanding the science of human behavior. Remember 2 per centers are the ones who find a way to *refocus* and accomplish their desired goal. *Which one are you; 98 per center or 2 per center?*

Number 1 cause of FAILURE in Profession and Life: INABILITY TO REFOCUS

> AN **INDIVIDUAL** WITH A **VISIONARY GOAL** THAT HAS THE ABILITY TO **MOTIVATE TEAM MEMBERS** TOWARDS ACCOMPLISHING
> # YOUR VISIONARY GOAL

CHAPTER 7 REFOCUS

SUMMARY

<u>Accomplishing Your Visionary Goal</u> *with Team Member Refocus*

All things of importance in one's life can be achieved with self-discipline

Developing *refocusing* ability is the most challenging of all the mental skills. Refocusing is elusive and requires the greatest degree of self-discipline. This is precisely why not *refocusing* is the number one cause of failure in leadership and in life performance. Individuals who have the greatest degree of successful *refocus* perceive a broken focus as an opportunity for a learning experience. Refocusing requires development of a *personal mental trigger*.

CHAPTER 7 REFOCUS: MAJOR TOPIC POINTS

Brief Discussion of Focus

Science: Knowledge of Refocus

Refocus: (Definition Why When How)

Refocus: The ability to regain maximum performance after a distraction

Why Refocus

Number 1 cause of Failure

Portrait of Refocusing

A Politician: Abe Lincoln

An Olympian: Greg Louganis

Portrait of Failure to Refocus

World Series: Kansas City Athletics vs. St. Louis Cardinals

When to Refocus

Before Performance

During Performance

After Performance

Art: Application of Refocus

How to Refocus

Mental Triggers in search of Alpha Brain Waves

Written Goal

A Word or Words

An Image

A Thought

A Touch

Ridiculous Association

Developing Your Personal Mental Trigger for Refocusing

98% or 2% which one are you

98% fail to Refocus

2 % successfully Refocus

CHAPTER 7 GLOSSARY: REFOCUS

Boxmentalize: Mentally, temporally putting away an incident to analyze at the appropriate time and making it a learning experience.

Devolution: Brain's development regressing

Focus: The ability to attend to proper stimuli during performance

Mental Trigger: Internal stimuli to direct thoughts in a deliberate positive direction

Refocus: The ability to regain maximum performance after a negative or positive distraction.

Mental Skills of Imagery, Relaxation, Refocus, are for the purpose of enhancing the successful Accomplishment of Goals

AN **INDIVIDUAL** WITH A **VISIONARY GOAL** THAT HAS THE ABILITY TO **MOTIVATE TEAM MEMBERS** TOWARDS ACCOMPLISHING **YOUR VISIONARY GOAL**

PART II:

SUMMARY

◆ ◆ ◆

SCIENCE AND ART OF MENTAL SKILLS

The Science of Human Behavior will Empower You and Your Team Members in Accomplishing Your Visionary Goal

Each mental skill presented, enables you to pick and choose the glove of best fit for your personal style of leadership. Mental Skills of *Goals, Imagery, Relaxation and Refocus* were presented in two major categories:

1. Science: Knowledge of the Skill
2. Art: Application of the Knowledge

Understanding Science and Art of each skill enables elevating your leadership to its highest sustainable level

CHAPTER 4 GOALS

Goals: A road map, a strategy towards achieving a desired destination. Not only does proper goal setting identify strengths and weaknesses, it also identifies purpose, ownership, and direction for leadership, for individual team members, for your team, and your personal life. Developing a goals program will be the foundation of your mental skills program. If at first you choose to do nothing else but implement a goals program, this in itself will provide the highest level of sustainable success. *In the future*: Should you desire to apply the additional skills of imagery, relaxation, and refocusing; these mental skills will accelerate successful goal accomplishment throughout your organizational structure.

ADDENDUM

The Human Brain: A general discussion of the human brain enables a better understanding of the mental skills of Imagery, Relaxation, and Refocus. Rewiring the human brain is easily accomplished with proper exercise.

CHAPTER 5 IMAGERY

Imagery: Creation of a mental image may include one or more of the five senses. Research indicates at least 80% of the same brain waves are used in imagery as compared to physically performing a skill. One of the major imagery benefits is an enabler for building self-confidence in pursuit of goal achievement.

CHAPTER 6 RELAXATION

Relaxation (*Alpha Brain Waves*): The ability to control one's stress level. Achieving flow, achieving the zone, being in rhythm is dependent upon your ability to control stress level to enter the Alpha Stage. In your professional and in your personal life developing relaxation skills is no longer a luxury; it is a necessity for sustaining your highest level of leadership success.

CHAPTER 7 REFOCUS

Refocus: Ability to regain focus after original focus has been broken by either a negative or positive circumstance. The ability to refocus requires developing a mental trigger. Developing the ability to refocus enables sustainability of success in profession and in personal life.

*The Number 1 cause of failure in Leadership and in Life: Inability to Refocus

MENTAL SKILLS LEADERSHIP FOR PROFESSION AND LIFE

SUMMARY

All roads to successful accomplishment are through the mind. All successes are first born and won there and all failures are first born and lost there. *West Point Military Academy*

Yesterday, the science and art of mental skills were limited to those who studied human behavior. Example: In 1995, as a college student, I gave a presentation on mental skills with emphasis on the ability to refocus. After the presentation a female colleague approached and commented, "Had I met you two years ago I would still be playing professional tennis." In application of *Mental Skills Leadership*, your knowledge becomes invaluable; not only to yourself, but also many of the lives you touch. When your leadership transcends mental skills throughout your organizational structure, the benefits in professional and personal life will be numerous and some instances life-saving.

Previously presented was the example of Greg Louganis hitting his head on the diving board in a crucial moment during Olympic competition. His ability to refocus allowed his next dive to be near perfect.

Following are examples of how each one of us have, or once again will hit our head in our personal life's journey.

LIFE CRISIS

Each one of the following examples requires the mental skill of *refocusing* in pursuing future success.

> *Divorce, Death, Loss Employment, Bankruptcy, Terminal Disease, Cancer, Crippling Accident, Return from War, Care Givers, Weather Devastation, Immigration, Betrayal, etc.*

Today the Science and Art of Mental Skills are much more prevalent than ever before. *MSL* presented a strategy to elevate your successful performance in Leadership and in Life to its highest sustainable level.

SUSTAINING YOUR HIGHEST LEVEL OF LEADERSHIP

The greatest leadership in all professions, including military leadership of every nation, is susceptible to poisonous traits of human error, affecting sustainability of success. Every individual who establishes leadership to a higher level has the mental capacity of sustaining their successful leadership.

Understanding the Science and Art of human behavior enables you to successfully address the many different negative issues. Satisfaction and Complacency are two of the most common human traits always available to destroy leadership. The following is an example:

A GREAT LEADERSHIPS FAILURE

World Renowned

In 1967 Israeli military forces entered and completed a major battle in the Middle East in six days. This incredible victory was so extraordinary it has become known worldwide as The Six Day War.

The excellence of these military leaders and soldiers became a standard for which other nations' militaries would measure themselves. However, looming in the shadows of darkness were the human traits of satisfaction and complacency.

While the Israelis were basking in the glory of success, their opponents were regrouping. Having suffered a major and humiliating defeat, Syria and Egypt regrouped and improved their training and battlefield strategy. In 1973, the two Middle East countries attacked and successfully crossed the Suez Canal. Syria and Egypt regained much of the Golan Heights.

Regardless of profession, regardless the level of leadership, regardless of personal life, satisfaction and complacency are toxic tenets of leadership with eventual result of failed sustainability. This is another challenge for leadership, which requires the science of human behavior. Balance of *What Am I with Who Am I* eliminates toxic tenets of leadership. Proper Goals combined with Refocusing enables sustainability.

Mental Skills Leadership for Profession and Life has provided a linear strategy for elevating your professional and personal leadership to its highest sustainable level without destroying the rest of your life.

MSL LINEAR STRATEGY

* Develop Your Philosophy of Life
* Develop Your Philosophy of Profession

* Systematically Evaluate Your Critical Balance - *What Am I with Who Am I*
* Develop Your Leadership Visionary Goal
* Develop Your Leadership Performance Goal
* Create a Motivational Environment for Your Team Members
* Counsel the Development of Team Member Destination and Performance Goals
* Implement Goals, Imagery, Relaxation, Refocusing for Accomplishing Your Visionary Goal.

An Individual with a Visionary Goal that has the Ability to Motivate Team Members towards Accomplishing Your Visionary Goal

AN **INDIVIDUAL** WITH A **VISIONARY GOAL** THAT HAS THE ABILITY TO **MOTIVATE TEAM MEMBERS** TOWARDS ACCOMPLISHING **YOUR VISIONARY GOAL**

MSL

DESTINATION AND PERFORMANCE GOAL

Mental Skills Leadership's Destination Goal: Present to you a successful strategy for elevating and sustaining the highest level of leadership while maintaining an enjoyable, healthy personal life.

Mental Skills Leadership's Performance Goal: Present to you a meaningful, applicable, thought provoking, enjoyable experience.

Those of you who may have suggestions or comments for *MSL*; we welcome your thoughts. leadershipmentalskills@gmail.com
When the Persian military officer Tigranes heard that the prize was not money, but a crown of olive, he could not hold his peace, but cried, "Good heavens, Mardonius, what kind of men are these that you have pitted us against? It is not for money they contend but for glory of achievement!" Herodotes, Histories, 8.26.3

REFERENCES

Asimov, I. (1991). How did we find out about the brain?. New York: Scholastic, Inc.

Bandura, A. (1977). Social learning theory. Upper Saddle River, NJ: Prentice Hall.

Cloud, J. (Nov. 14, 2011). Thought control, Time Magazine. Retrieved from http://www.time.com/time/magazine/article/0,9171,2098588,00.html

Corey, G. (2011). Group counseling. Pacific Grove, CA: Brooks/Cole Publishing Company.

Cox, R. H. (2011). Sport psychology concepts and applications. Dubuque, IA: Wm. C. Brown Publishers.

Goleman, D., Boyatzis, R. & McKee, A. (2004). Primal leadership: realizing the power of emotional leadership. Boston, MA: Harvard Business School Press.

Greensburg, J. S. (2010). Comprehensive stress management. Dubuque, IA: Wm. C. Brown Publishers.

Hardt, J. (2005). Alpha waves-alpha brain waves. *Advanced Brain Wave Biofeedback Training.* Retrieved from http://www.biocybernaut.com/about/brainwaves/alpha.htm

Hayward, S. (1998). Relax now removing stress from your life. New York: Sterling Publishing Co., Inc.

Hunt, M. (1983). The universe within. New York: Touchstone Simon & Schuster.

Johnson, A. G. (2000). The Blackwell dictionary of sociology. Malden, MA: Blackwell Publishers Inc.

Kotulak, R. (1997). Inside the brain. Kansas City: Andrews McMeel.

Lynch, J. & Chungliang, A. H. (2006). The way of the champion. Rutland, VT: Tuttle Publishing

Martens, R. (1997). Coaches guide to sport psychology. Champaign, IL: Human Kinetics Publishers, Inc.

Maxwell, J. C. (2011). The five levels of leadership: proven steps to maximize your potential. New York, New York: Center Street, Maxwell Book Group.

Millman, D. (1979). The warrior athlete body, mind, and spirit. Walpole, NH: Stillpoint

Publishing.

Murphy, S. (2004). The sport psych handbook. Champaign, IL: Human Kinetics Publishers, Inc.

Orlick, T. (2007). In pursuit of excellence. Champaign, IL: Leisure Press.

Steele, D. (Producer), & Turteltaub, J. (Director). (1993). *Cool Runnings* [motion picture]. United States: Walt Disney Pictures.

Sterner, T. S. (2005). The practicing mind. Wilmington, DE: Mountain Sage Publishing.

Sullivan, G. R. & Harper, M. V. (1996). <u>Hope is not a method.</u> New York: Random House.

Williams, J. M., ed. (2009). <u>Applied sport psychology.</u> Palo Alto, CA: Mayfield Publishing Company.

Glossary

Additive Principle: The notion that intrinsic and extrinsic motivation combine to create need achievement.

Alpha Brain Waves: Present in wakefulness where there is a relaxed and effortless alertness. Alpha waves oscillate 8-13 times a second.

Art: The application of knowledge

Boxmentalize: Mentally temporally putting away an incident to analyze at the appropriate time and make it a learning experience.

Carpe diem: Seize the day

Coaching: Ability to UNDERSTAND the heart and mind of team members thus enables identifying team member's strengths and weaknesses.

Competence: Competence is a team member's conviction for a successful outcome

Counseling: Issue and conflict resolution solved by GUIDING the team member to discover the solution

Creative Imagery: Individuals with advanced imagery are able to create visuals and to create motion (usually a performance) then store these created pictures and movements in the memory for future use

Crisis: An unfortunate happening which requires the ability to refocus for a positive outcome

Dimension: A new measurement

Effective Communication: When you the sender create a duplicate picture in the mind of the receiver.

Empowerment: Leadership providing team member with necessary physical, technical and mental skills for successful performance within positional boundaries

Environment: A carefully crafted atmosphere that provides boundaries from which a team member may pick and choose from a selected menu of acceptable stimuli (mental techniques) designed for a specific expectation which influences behavior and outcome

Extrinsic: Motivation that comes from an outside source. External rewards to motivate behavior in performance such as, money, ribbons, medals, trophies, and praise

Failure: Not a success, disappointment

Feedback: Leadership's <u>counseling</u> comment on team member performance immediately or soon after performance. (Verbal and recorded)

Flow: A euphoric mental state of an effortless highly successful performance

Focus: The ability to attend to proper stimuli during performance

Galvanic Skin Response (GSR): Skins electrical field causing minute muscular activity which may be controlled by ones thought process.

Hubris: Drunken with power

Imagery: An imagined visual imitating reality which serves a purpose

Insecure Leader: Fundamentally insecurity is fear. Leader requiring having their ego puffed up by as many team members that can be convinced, coerced or threatened to do so

Intrinsic: Motivation that comes from within an individual (team member) Motivation to participate in an performance for its own sake and for no other reason

Leadership's Performance Goal: Your strategy for the accomplishment of your visionary goal

Leadership's Visionary Goal: Your vision for a targeted outcome within a specified time

Mental Trigger: Internal stimuli to direct thoughts in a deliberate positive direction

Motivation: The degree of mental intensity directed towards the accomplishment of a goal

Multiplicative Principle: The belief that intrinsic and extrinsic motivations are interactive and not additive

Nepotism: Favoritism; Bias

NREM (Non Rapid Eye Movement): One of the brains cycle during sleep. Dreams during NREM are believed to be pieces from the past.

Parasympathetic Nervous System: Assists in returning the human body to a relaxed state after experiencing increased energy.

PED: Performance Enhancing Drugs

Philosophy of Life: All that is acceptable within a personal boundary (may include purpose and direction)

Philosophy of Profession: Perception of your purpose in your profession

Psychoneuromuscular Theory: This scientific theory suggest imaging certain actions produces the same path for neuromuscular messages that are transported via the right hemisphere of the brain to the correct muscles. A psychoneuromuscular response providing an individual a more automatic response in the same or like situation

Public Criticism: Public Denigration

Recall Imagery: Individual recalling visuals which may also include movements and senses from the brain's memory.

Refocus: The ability to regain maximum performance after a negative or positive distraction

Relaxation: Cognitive control of stress level

REM (Rapid Eye Movement): One of the brains cycle during sleep. Dreams during REM are believed to be pieces creating the future.

Revenge: Reprisal; Pay Back

Science: The knowledge of

Secure Leader: Self-confident leader keeps their ego in check for the benefit of improved performance by each team member

Self-Discipline: Personal Control

Stair Step Method: Strategy for developing team member Destination Goal

Stolen Identity: When the Who Am I is consumed by the What Am I

Success: Victory, accomplishment

Symbolic Learning Theory: Creates a "coding" of certain actions into a symbolic element allowing a more automatic response in like circumstances

Sympathetic Nervous System: Function is distributing and increasing energy (stress); causing the elevation of blood pressure, increased breathing, heart rate, perspiration

Team: *E Pluribus Unum* "Out of many, one"

Team Culture: Shared norms and values

Team Member's Destination Goal: Team member's targeted outcome within a specified time relating to successful accomplishment of your visionary goal

Team Member's Performance Goal: Team member's strategy within positional boundary for successful pursuit of performance.

Visionary Goal: An imagined possibility, stretching beyond today's capability, providing an intellectual bridge from today to tomorrow

What Am I: Your profession

Who Am I: Your character